Challenging Legitimacy at the Precipice
of Energy Calamity

Debra J. Davidson · Mike Gismondi

Challenging Legitimacy at the Precipice of Energy Calamity

 Springer

Debra J. Davidson
Faculty of Agriculture
Life and Environmental Sciences
University of Alberta
Edmonton, Alberta T6G 2H1
Canada
Debra.Davidson@ales.ualberta.ca

Mike Gismondi
Centre for Social Sciences
Athabasca University
Athabasca, Alberta T9S3A3
Canada
mikeg@athabascau.ca

ISBN 978-1-4614-0286-2   e-ISBN 978-1-4614-0287-9
DOI 10.1007/978-1-4614-0287-9
Springer New York Dordrecht Heidelberg London

Library of Congress Control Number: 2011934248

Printed on acid-free paper

Springer is part of Springer Science+Business Media (www.springer.com)

(T)hey saw in each other's eyes the dread, the abrupt tearing sensation of doubt ... the superintendent had a vision. He saw like an opened book the immense curves of the Athabasca River swinging through wilderness down from the glacial pinnacles of the Rocky Mountains and across Alberta and joined by the Berland and the McLeod and the Pembina and the Pelican and the Christina and the Clearwater and the Firebag rivers, and all the surface of the earth was gone, the Tertiary and the lower Cretaceous layers of strata had been ripped away and the thousands of square miles of black bituminous sand exposed, laid open, slanting down into the molten centre of the earth, O Miserere, miserere...

"The Angel of the Tar Sands" by Rudy Weibe in Rudy Weibe, *The Angel of the Tar Sands and Other Stories,* Toronto: McLelland and Stewart, 1982. Originally published in Rudy Weibe 1979 Alberta/A Celebration, Edmonton: Hurtig Publishers.

Dedicated to William R. Freudenburg and William Fuller, who left us before they were done.

Preface

Popular discourses on looming catastrophes of all sorts tend to fall into two patterns: either we are doomed and there is nothing we can do about it or "the system" will fix itself and there is nothing we need to do about it. Both are, in a sense, excuses for ignoring alarm signals, and avoiding pro-active planning for change, in effect shielding complex problems from critical examination and reflection. Regardless of one's interpretation of history, however, one would be hard-pressed to identify instances in which avoidance ever amounted to positive outcomes. This book is an attempt to dig deeper into just how avoidance becomes legitimated, even in those cases when doing something urgently would seem quite prudent. The case we focus on is the development of bitumen in northeast Alberta, Canada, known as the Athabasca tar sands, or more colourfully "Dirty Oil" to critics, and oilsands, or most recently "Ethical Oil" to proponents. The Athabasca tar sands has attracted the attention of concerned citizens across the globe. In response, state and industrial interests have made significant investments in research and technological innovations intended to "green" the bitumen extraction process, and even larger investments in discursive framing to assure observers that such technological innovation and new scientific knowledge will avert ecological catastrophe, among several other conceptual frames intended to divert or otherwise marginalise critical attention. Such efforts at reframing have met with mixed success. Many critics view such manoeuvers as efforts to justify the massive and destructive expansion of the tar sands for the purposes of generating wealth for the few, leading to a political dance played out on the media floor, in hearing rooms, and on the internet. Still other ramifications of non-conventional fuel development are entirely absent from political discourse, despite very good reasons for attention to them. The outcomes of such discursive theatre have at least as much to do with our present and future relationship to energy as do the activities of world leaders and petrochemical geologists, and are worth watching closely.

The Athabasca tar sands has received a significant amount of attention by journalists, politicians, and social movement activists, but relatively little from the social sciences. We consider this gap in urgent need of rectification, and hope that our contribution encourages further critical inquiry from the academy. While the Alberta tar sands could be seen as just one of many industrial developments with harsh environmental consequences taking place across the globe today, and one that is in a rather remote geographic area that has not been subject to extensive global political attention previously, we consider this enterprise to have much deeper implications. Non-conventional fuels are those which require significantly larger amounts of inputs – in terms of raw materials, energy, labour, technology, processing, and so on – in order to transform them into a form that can be used. They are also, as one might anticipate, associated with much higher levels of waste and environmental degradation. To date, non-conventional fuels have not been developed extensively due to the relatively large input costs. More recently, however, increasing oil prices, combined with technological developments, and the precipitous decline in discovery rates of new conventional sources have drawn attention to the globes' several deposits of heavy oil, shale, bitumen, and deepwater reserves. Due in no small part to the perseverant advocacy of the Provincial state of Alberta, the Athabasca tar sands is the first large-scale non-conventional fossil fuel development that has become a significant producer of oil for the global marketplace. What happens here in the Athabasca tar sands could determine whether it is the first of many more such developments to come, or the last.

This book will contribute to the study of environmental and natural resource politics, and to the sociology of language and power, but on a much broader level, this book is about social change, a topic of urgent contemporary interest both within the academy and beyond. How modern social systems respond to our rapidly disintegrating relationship with easy energy can tell us much about the potential for collective agency on a macro-scale. As dictated by the context in which energy development occurs, we embrace contemporary conceptualizations of society as a complex system of flows and mobilities, rather than static entities, and yet simultaneously highlight the extent to which modernity is indelibly rooted in place, with both material and ideological implication. Within this turbulent system – as is the case in any dynamic, complex system defined by unpredictability and interconnected networks – lay sources of crisis, but also hope for positive societal transformation.

Edmonton, AB   Debra J. Davidson
Athabasca, AB   Mike Gismondi

Acknowledgements

As with all intellectual endeavours, we have many people to thank for their contributions to this book. Several of our colleagues have contributed simply (but nonetheless significantly) by offering their ears to our many ideas, and their honesty regarding the merits of those ideas. Others have contributed in more substantive ways, by assisting with data gathering and management, including Barbara DeRossi, Kendra Isaac, and Andriko Lozowy, all former graduate research assistants at the University of Alberta, and Andrew Anderson, undergraduate research assistant, also at the University of Alberta, and Michelle Campbell and Geoff Loken at Athabasca University. In addition, professional editor Lori-Ann Clarehout offered exceptional editorial advice, always with the utmost diplomacy. Finally, our gratitude extends to the Canadian Social Sciences and Humanities Research Council, who provided funding for this project.

Contents

Chapter 1
Look Who's Talking

Human history has often been described as a progressive relinquishment from environmental constraints. Now, it seems, we have come full circle. The ecological irrationalities associated with industrial societies have a lengthy history, and our purpose is not to catalogue this litany of horrors. Collectively, however, we have crossed certain literal and figurative thresholds. Our ability to commit such acts of irrationality unconsciously, obliviously, or simply nonchalantly, is no longer a luxury we enjoy. Two intersecting moments define this historic nexus. The mounting evidence for global climate change, now unequivocally attributed to socio-economic activities, has accumulated to the extent that today even insistent deniers must concede defeat in any but the most closeted of social circles. Simultaneously, as fossil fuel seekers come home with ever-shrinking finds, the end of easy oil and the easy wealth it has generated is upon us, rendering non-conventional fossil fuels the next most attractive option to those states and corporations that control the extraction and delivery of energy resources. The development of such non-conventional fuels does nothing to alleviate either climate change or peak oil, however. On the contrary, the tremendous energy requirements for extraction and processing of these low quality fuels translates into an abysmal and ever-decreasing return on energy investments, escalating greenhouse gas emissions, and other environmental disruptions. These two facts—growing reliance on non-conventional fuels and increasing scale of environmental disruption—are not merely co-incidental: as "easy" global reserves of fossil fuels in general, and oil in particular, become depleted, the remaining fossil fuels are far more difficult to source, and hence environmentally, socially and economically more costly to extract (Klare 2008).

The Future Is Here

The largest of these endeavours – indeed the largest industrial project in history – is taking place today in a remote land that until recently was home to a greater number of caribou than humans. Various geological processes several millennia in the making

D.J. Davidson and M. Gismondi, *Challenging Legitimacy at the Precipice of Energy Calamity*, DOI 10.1007/978-1-4614-0287-9_1,
© Springer Science+Business Media, LLC 2011

Fig. 1.1 Open Pit. Syncrude Aurora North Mine with permission of Louis Helbig. B2402051. See
http://www.beautifuldestruction.com and http://www.louishelbig.com/

have deposited a gargantuan volume of bitumen, more descriptively known as tar or
oil sands, into the sandy terrain lying below the boreal forest floor of north-eastern
Alberta. While the region's original peoples viewed this substance in the most utilitarian of manners – as an effective sealant for canoes – the few European travellers and
their descendants who first ventured into the area expressed a level of veneration for
the black sand often reserved for religious figures. Throughout much of the history of
European occupation, the emancipatory dreams imbued in Alberta's tar remained just
that, but the scale of tar sands development in recent years has surpassed the wildest
imaginations of early explorers. Today, euphoria about the Alberta tar sands is shared
by dozens of the largest, most profitable corporations in the world and tens of thousands of migrant workers who have descended upon Alberta's backwaters to strike it
rich or enjoy the ready pleasures that only hard cash can buy. The Alberta Energy and
Utilities Board (EUB) estimates that northern Alberta holds approximately 1.7 trillion
barrels of crude bitumen, although only 174 billion barrels are considered recoverable
using today's technology under current economic conditions. Existing production
averages around 1.5 million barrels per day (mbd), but anticipated output is expected
to reach anywhere from 5 to 10 mbd by the 2030s, involving some 33 mines (Fig 1.1),
83 deep-well drilling projects, and 40 upgraders that have been announced, applied
for, or are currently operating (Dunbar 2009). What do we get from one barrel of
synthetic crude oil? About enough gasoline to fill the tank of an American car.

Other observers have been captivated by a different symbolic representation of
the industrial-scale development of the Alberta tar sands, as a significant *threshold*

beyond which the ecological and social costs of development outpace the benefits received by humanity. International attention has been focused on the approximately 30 megatons (MTs) of greenhouse gases that are currently emitted from the Athabasca tar sands operations, projected to increase more than tenfold to 400 MTs by 2050 – in part a result of the high energy requirements for tar sands extraction, which are currently supplied by another fossil fuel, natural gas (CAPP 2009; Alberta Environment 2009d).[1] The contribution to global warming is by no means the only form of environmental disruption, however. Approximately half of the bitumen is extracted by mining, which to date has created open pits covering 530 km^2 (Alberta Environment 2009c). The other half is derived through deep-well drilling processes that consist of injecting steam underground to make the deeper tar deposits viscous enough to be pumped to the surface, entailing considerable habitat fragmentation that differs from mining, but is no less significant, and with much larger energy input requirements. Moreover, the extraction process is water intensive, consuming between two and five barrels of water per mined barrel, and around half a barrel of water per deep-well barrel produced (Alberta Environment 2009a). All contaminated water not fit for re-use is deposited in tailings ponds, which currently cover 130 km^2 of land (Alberta Environment 2009b).

These disruptions describe certain central, and yet often unheeded material flows in our contemporary *Liquid Modernity* (Bauman 2000), material flows that belie and yet intersect the far more rapid movements of people, ideas, and capital. As few of these flows are subject to the control of states, the power of states to draw together their citizens as a unified body is weakening, as is their coercive power (Urry 2000). Sources of credibility, culpability, responsibility, and legitimacy have all been de-stabilized, becoming ambiguous, manipulable, and ephemeral. This is not to downplay the persistence of certain forms of power, but rather to suggest that avenues of power's expression are evolving: as with other features of social systems, power itself becomes better understood as *flow* than structure. At the heart of this liquid modernity are global communications, bringing the formerly distant into close range, enhancing its tangibility. The fluidity of communications has escalated through digitization, compromising the ability of states and other interested parties to control flows of ideas and information, which have become formless, chaotic, and unpredictable. Digital media pitches the public world into the private, and the non-local into the local, abolishing the perceptual and sensory distance between oneself and faraway places, events, and people (Urry 2000). Optimistic researchers associate global communications networks with the prospect of a global civil society (Rheingold 1994), in part by introducing opportunities for an ever-growing cadre of others to question and intervene in previously exclusive political discourses. Emergent information and communication technologies have unquestionably contributed to the

---

[1] To recover one barrel of bitumen by in situ extraction, 1,000 cubic feet of natural gas is required, and 250 cubic feet of natural gas is needed for extraction by open-pit mining (Alberta Chamber of Resources 2004). These figures do not include the energy consumed in the upgrading, refining and transport of fuels.

proliferation of information sources, provided new forms of participation, and created new opportunities for re-establishing relationships between citizens and politicians (Bentivegna 2006). One should not wax too optimistic about the transformative power of global communications, however, as multiple peoples have no participatory access to this discourse at all, and certain forms of political power remain quite entrenched.

One of the primary reasons certain forms of power are difficult to unseat is the close relationship between political power and control over another set of global flows that are exceedingly material, namely earth's resources. While Urry and others chide sociology colleagues for failing to grasp a "sociology of flows", social scientists working in certain sub-disciplines have been employing a sociology of flows of raw materials for decades, beginning with Innis (1956, 1940). Innis' staples theory described how flows of natural resources from hinterland to core shaped the social structures of both – flows directly defined by transportation and communication channels that underwent continual change. Several geographers have donned a similar lens, following Massey's (2005, 1995, 1994) lead in tracing the spatial flow and distribution of capital and labour. More recently, this tradition has emerged in the form of global commodity chain analysis (e.g. Gereffi and Korzeniewicz 1994). The late Stephen Bunker, both in his earlier work (Bunker 1985) on global raw materials and ecosystems and recent collaborations with Paul Ciccantell (2005), has provided one of the most insightful frameworks for analyses of material flows since Innis. Bunker argues that the very impetus towards globalization of our social systems is driven by capital's imperative to reduce production costs by minimizing raw materials inputs. But historic instances of success at doing so have only provided incentives for expansion of production – another inherent capitalist imperative – leading to a net increase in raw material use and ecological impact, and the search for new reserves that are frustratingly contained in ever more distant locations. In contrast to many contemporary portrayals of global capital processes as spatially and materially unbounded, "it is impossible to pretend that raw materials deposits, mines, railroads, and steel mills are footloose" (Ciccantell and Smith 2009:362). This body of work, in a nutshell, emphasizes the historic and very material precedents to our contemporary fluid modernity, and put to rest any sentiments regarding the "de-materialization" of contemporary societies. These critiques of the de-materialization thesis provide a nice complement to Urry's sociology of flows, by problematizing the material and ecological sides of the equation.

Material flows are part of the history of civilization; however, the exponential increase since the 1970s in the mobility of people and information in time frames ever closer to instantaneous marks an exponential shift in the complexity and dynamism of modern society. Today's global map overlays a pre-existing global economy founded upon the flow of remote physical materials that were highly resistant to mobility. Seemingly immaterial, the emergent network of flows has ironically increased pressures on material flows, while simultaneously masking those material realities, as shifts in the paths of material dispersion are readily mistaken for ecological modernization and de-materialization. Outcomes are not predetermined; however, as the same processes enabling escalation also provide avenues for *revelation*, simultaneously enhancing the need for, and opening avenues for, greater reflexivity. Such reflexivity is expressed in the form of increased awareness of ecological crises, interpretive struggles over the meaning of such crises, and expression of social discontent, all with unpredictable end points.

One critical ramification of Liquid Modernity is clearly expressed in this current context: the option for personal dissociation – however removed one is by geography or awareness – is lacking, rendering complicity, conscious or otherwise, among all members of global society. Consumers do not have a choice whether or not to consume energy after all. Even the presumed choice *among* energy supplies is fallacious. While boycotts of "dirty oil" are certainly politically meaningful, the choices posed are merely a facade when applied to global supplies of energy staples. An individual, community, or state that refuses to purchase petroleum products from the tar sands and yet replaces them with equal amounts of fossil fuels produced elsewhere does nothing to reduce demand on a shrinking pool of global reserves, and inevitably that demand increases exploitation of those same non-conventional fuels. The futility of our convenient subsuming of environmental concern into a green marketplace is laid bare; our choices among blood diamonds and the "guilt-free" diamonds of Canada, between Nike and Birkenstock, between conventional and organic foods are nearly as speculative. If we choose to continue to support a dominant global economic structure premised on "fossil capitalism" (Huber 2008), the necessarily escalating reliance on non-conventional fossil fuels will come at a huge cost to social and ecological wellbeing. The scale of those costs demands recognition and attribution, and the freedoms enabled by fossil fuel consumption come with a responsibility to deal with irreversible consequences; ultimately "more freedom means less choice" (Shove and Warde 1998:7). As we will argue, these freedoms and responsibilities raise the obvious questions: What are the likely avenues of response by societies? Without the luxury of naiveté, does not a collective agreement to engage in suicidal tendencies have foreboding implications for social order and human psyche alike? The only way forward – if forward movement is at all plausible – is with our eyes wide open. Will the Athabasca tar sands become a global wakeup call? Our twenty-first century Titanic?

Where Do We Go from Here?

If the potential for collective response to threat exists at all, one would think that potential would be ripe now. This potential, however, often receives little scrutiny; most commentators presume that such a response is either determined by structural conditions or entirely unrealizable. Setting aside the latter group of fatalists, the former group tends to be divided into three categories. First, many pull out that now age-old, reliable calling card, technological optimism (Lomborg 2001; Simon 1981). Technology has without question enabled seemingly miraculous changes in our relationship with the biosphere throughout history, so why should the tremendous potential of human ingenuity be doubted now? A role for technology in responding to material and environmental crisis can certainly not be ruled out. But the exponential increase in resource supply and effort that would be required for the development of new technologies capable of accommodating an infinite growth trajectory cannot be ruled out either, even assuming solutions are possible within the confines of universal physical and ecological laws (a mighty lofty assumption indeed). States cannot simply continue to underwrite the costs of technological optimism through public resources, and markets will only do so to the extent that costs can be imposed on society in commodity pricing. But when the commodities in question include food, water, and energy, the "if they want it bad enough they will pay" clause embedded in market logic poses a serious ethical problem.

Others invest hope in the authority of states to intervene. But *why would they?* Even setting aside issues of state autonomy and capacity, *why* would a given state invest resources in substantive, as opposed to purely symbolic, responses to ecological degradation? Some presume a response in instrumental terms, like the much-anticipated internalization of ecological rationales into efficiency-based decision-making hypothesized by Ecological Modernization Theory (e.g. Mol and Sonnenfeld 2000; Spaargaren and Mol 1992). Others suggest states will assume responsibility for environmental well-being because *World Society* told them to (Frank et al. 2000). The historical record suggests otherwise. States are faced with limited sustainability imperatives, and more often than not are compelled to partake in just the opposite. To the extent that they are beholden to the corporations benefiting from environmental disruption, states have a structural incentive to avoid, rather than address, environmental problems. Of course growing concern in domestic and global civil society creates tensions for state actors, but fortunately for state agents facing such a bind, the direct *indications* of environmental degradation are often obscure, and in many instances dependent upon scientific interpretation by a techno-scientific elite, themselves divided and influenced by social pressures (Finlayson 1994). This set of circumstances contributes to an extraordinarily high degree of interpretive flexibility, such that interpretations themselves often determine political outcomes (Freudenburg et al. 2008), which have little more than symbolic influence on the condition of our ecosystems.

We are left then with reliance on the transformative potential of civil society. Recent encouraging signs are changes in individual-level environmental awareness and behaviour, and the growing political influence of environmental mobilizations of various forms. All too often, however, these efforts have been too little and too late. Even the ecologically conscious among us in the Western world face limited abilities to remove ourselves from our consumptive lifestyles. A society-led transition to a post-carbon society may well test the organizational capacities of civilization, but we warrant that is a test well-worth critical consideration. Several crucial preconditions would need to materialize. Unleashing this transformative potential requires first and foremost a legitimacy crisis to unfold, which in turn depends upon a particular reflexive course: the collective recognition of the illegitimacy of current forms of order and distribution; the imagination of alternative trajectories; and the predilection and capacity to act on such projects, by embarking upon untrodden and therefore uncertain future paths.

Legitimacy in Fluid Modernity

Centre stage in the current analysis is consequently taken by the role of legitimacy – described roughly as concession to the "justness" of given power structures, projects, and ideologies by those subjected to them. Legitimacy presumes a relation, in which some entity exerts power over another, and the latter concedes to such an imposition, forsaking one's own power to reject or rebel. While theoretical interest in legitimacy enjoys a lengthy scholarly history, the concept remains elusive. More often than not legitimacy tends to be presented in absolute terms: legitimacy is either present or absent. Yet since legitimacy is inevitably embedded in fluid modernity, it is more appropriately measured in degrees rather than absolutes, and as a flow rather than a structural condition. Likewise, the social contract itself and its component parts are culturally and historically contingent. Both the expectations on states associated with that contract and the delineation of societal groups to which the contract pertains are quintessentially fluid, defined by globalized markets and polities, creating an ever-changing cartography of routes through which legitimacy crisis may emerge (Blanchard et al. 1998; Nye and Myers 2002).

Empirical researchers usually resort to measurement of hypothetical indicators of legitimacy's presence. The *presence* of conditions favouring legitimacy, such as shared beliefs, is by necessity conjoined with certain notable *absences*, however, in that any given perspective is inevitably partial, concealing as much as it promotes (Freudenburg and Alario 2007). As Habermas (1975), among others, has been wont to point out, dominant ideologies serve to support the inequitable distribution among its subjects of both wealth and threat. Inequities are not the only form of contradiction that must remain concealed, however, in order for legitimacy to persist. Some belief systems are also associated with deeper structural irrationalities, the revelation of which would be equally threatening to those who benefit from the current social order. Steadfast ascription to continuous progress from within the confines of an economic system wholly dependent upon finite resources and waste sinks is one such irrationality. The treatment of these same sources and sinks as isolated from the biosphere in which they exist is another. While the application of the concept of legitimacy is often restricted to institutions of authority – namely states – legitimacy is tied not just to the stability of authority structures but also to the ideologies themselves that states inevitably serve to endorse, and the particular projects that embody those ideologies. If legitimacy is enabled by the presence of shared beliefs, it stands to reason then that it is the presumed legitimacy of those beliefs themselves that, by extension, affords legitimacy to the authority structures and projects embodying those beliefs.

Neither inequity nor irrationality need pose a threat to legitimacy, however, provided the indications of such contradictions either remain concealed or can be effectively explained away. One main effect of political discourse, in addition to the promotion of both its own authority and the ideologies espoused, is to *conceal or discount* the inequities and irrationalities endorsed by state and ideology alike. Legitimacy is thus sustained only to the extent to which such contradictions remain unacknowledged. Such legitimacy is enhanced through the control of information through certain mechanisms – including the mass media in particular – which have the effect of maintaining the hegemony of certain beliefs.

Energy provides a particularly complex and useful venue for evaluating this fluid legitimacy. It is an enterprise in which, while benefits are most certainly not distributed equitably, we do all nonetheless benefit. Or more to the point, we would all (and many do) very much suffer without it. At the current nexus, however, the level of scrutiny to which states are exposed continues to escalate, suggesting the possibility that activities un-noticed in the past can generate significant controversy for state and corporate proponents of certain forms of energy development today. The rising potency of global warming and peak oil as political (and economic, health, security, etc.) issues have broadened the parameters of critical scrutiny, raising the spectre of legitimacy threat to the realm of global market share and international reputation. The growing disruptions caused by the development of our remaining oil reserves may well pose a legitimacy challenge for states today, even within petro-states that have historically enjoyed high levels of local support, or at least quiescence. At the same time, those states with jurisdiction over the last remaining reserves of fossil fuels may face increasing economic incentive but also political pressure to *develop* those resources, ostensibly to avoid an energy crisis, placing state actors in a rather delicate position on the domestic and international stages, in which discursive manoeuvrability is key to the maintenance of legitimacy.

What is interesting here is the prospect of persistent adherence to the legitimacy of certain social structures to exacerbate the likelihood for contradictions to arise, and undermine the very foundations of those social structures. Beck (2009) argues that the accumulation of hazards at the global scale will transpire in reflexive modernization, a process of learning from and responding to structural disfunctionalities at an institutional level. Unfortunately, Beck presumes that the experience of collective victimhood will in and of itself transpire into societal response. For us, any hope for social transformation beyond both peak oil and global climate change rests in the potential for a sufficient number[2] of individuals to call into question the legitimacy of those social contexts, and secondly to pursue courses of action – including collective action – that are intended to encourage the transformation of those social contexts. Legitimacy crisis, in other words, does not rest solely on the acknowledgement of contradictions. Legitimacy cannot be conceived of as a deterministic source of agency, the absence of which automatically triggers social struggle. Many social structures clearly lacking in legitimacy as it is defined here have persisted for great lengths of time. A legitimacy crisis requires individuals to express their discontent, and design and pursue projects intended to establish new norms and rules of behaviour.

---

[2] What that number is poses a frustrating source of elusion for empirical researchers, but conceptually it is determined by the level of accumulated behavioural change that is needed to overwhelm the absorption capacity of the existing social system.

Discursive Channels of Fluid Legitimacy

The question then becomes, to where does the social scientist turn for evidence of such reflexivity? It is not contradictions themselves, but the interpretations of those contradictions, that direct avenues of response. We can learn a great deal about legitimacy, then, by exploring the ongoing discursive practices employed by political agents as a means of negotiating or contesting legitimacy. Many evaluations of the causes of and responses to environmental degradation tend to focus on certain structural and/or objective conditions as explanatory factors, some favourites including capital, technology, and population. A small but compelling literature on political discourse offers a more nuanced contribution, however. Analysis of discourse in environmental and natural resource politics holds tremendous intellectual merit, given the dynamic intersections between material and non-material, objective and subjective matters that both highlight the fluidity of modernity and ground that modernity in physical space.

All Eyes on Alberta

The rapid pace of expansion of the Athabasca tar sands over the past decade can be seen as a moment of dislocation, during which long-standing material and discursive regularities and routines that were previously sufficient have been subject to challenge; a time at which development proponents must mobilize new discourses and connect the previously unconnected in order to defend their legitimacy (Hajer and Versteeg 2005; cf. Howarth 2000). Processes characterizing the development of the Athabasca tar sands, and the legitimation of that development, are deeply integrated into a global network of flows. But equally pronounced are the particulars of space and directionality, describing a map that is in many ways rigid, consisting of relatively resilient patterns of material and discursive flow.

On Space

Petrochemical products circulate the globe in large quantities, yet all such resources must come from somewhere, and go to somewhere. Burawoy et al. (2000), among others, caution against conceptual frameworks that *over*-emphasize flows at the cost of space. Those spatial locations, and the extent to which such resources in their original form are physically "attached" to those spaces, play a fundamental role in the form that development takes, and the resulting impacts. First, the spatial separation between frontier developments – such as the Athabasca tar sands – and population centres has historically tempered public reaction to the land-based impacts of such developments. This is certainly the case in Canada, in which the national population is heavily concentrated in urban centres within a few hundred kilometres of the US border, while most such developments take place much further north. The very spatial extent of the country, in contradistinction to the geographic concentration of its peoples, has nourished an "unlimited frontier" imaginary among Canadians well into the twenty-first century – tempering concern for land disruption. At the same time, for the consumer, those natural resources come from so far away that they effectively become de-linked from a particular place, thereby relieving consumers of responsibility for production's impacts (Princen et al. 2002).

In addition, resources that are fixed in space are subject to tenure, and the institutions with political jurisdiction over those spaces have a disproportionate degree of power to dictate the pace and character of development trajectories. While many social-scientific treatments of development – indeed most sociological analyses across the board – utilize the nation-state as the preferred unit of analysis, such a selection would be inappropriate in the current context. With a federalist political structure in which a tremendous amount of jurisdiction over land and environmental concerns remains at the provincial level, the Canadian nation-state offers an inappropriate lens for analysis of tar sands development. The Province of Alberta, rather than the Canadian federal government, has historically been not just the most active political body associated with tar sands development, but also officially has the greatest level of authority to do so. And according to some, this provincial state enjoys a tremendous amount of political weight: as one recent commentator put it, "the province [Alberta] can control its own destiny more than any other because, in the years to come, Canada will need Alberta far more than Alberta will need the rest of Canada" (Maich 2005). While such political power has without question been enabled by the spatial presence of the resource itself, it has not been simply bestowed upon the Province directly, since the potential power embodied in the resource does not emanate from its mere existence, but rather in its potential for commodification.

To this end, the Province of Alberta has been actively – even aggressively – encouraging tar sands development for nearly a century. Without significant levels of state endorsement and fiscal subsidization the tar sands would never have been developed, nor would such development be able to persist (Gillmor 2005). Provincial government administrations since the first Premier took office in 1905 have prioritized the development of the tar deposits, by engaging in close relations with industry, supporting research in Alberta's universities and in the Province's own Alberta Research Council, and providing direct fiscal incentives to encourage private investment. For over 3 decades before the first commercial barrel was produced, the Province had been a primary sponsor of the research necessary to generate the technological developments that enabled the commercialization of the Athabasca bitumen (Chastko 2004). Today Alberta could be described in many respects as a prototypical neo-liberal Petro-state (Shelley 2005), with the energy industry contributing an average of 42% of provincial GDP, 31% of provincial employment, and 32% of government revenues over the period of 1971–2004 (Mansell and Schlenker 2006). On the other hand, to understand the revenues the state receives from the sale of its resources simply as "rent", would be misleading. This rent comes in exchange for substantial (albeit many would argue inadequate) state subsidization of development, including decades of research prior to and since commercialization, all manner of physical infrastructure needed to traverse the extensive spaces between sites of production and consumption, exclusion of a substantial land base from other forms of development, and the social service demands of a boom economy.

The power of the Albertan state to influence the development of the tar sands is mirrored by the *potential* political power of Alberta's citizens relative to citizens elsewhere. This is the case in part because Alberta's citizens formally wield the greatest power to remove the legitimacy of the Albertan state, in the voting booth. Secondarily, however, the relative distance between the tar sands and the vast majority of members of global civil society minimizes the visceral and experiential reaction among members of that non-local citizenry that might engender mobilization, rendering local citizens the more likely source of sustained opposition by comparison. Albertans, however, have not engaged in substantive levels of organized opposition to date. To the contrary, voters have expressed consistent support for the Progressive Conservatives – the provincial party that has been the most outspoken advocate for the development of the tar sands, an enterprise that has furthermore been pursued according to strictly neo-liberal ideological guidelines. The Progressive Conservatives actually gained seats during the most recent election, at a time when tar sands development was at its most acrimonious.

The potential for local opposition to the actions of tar sands proponents is more accurately reflected in trends other than election outcomes, however. Provincial elections have been plagued by persistently low voter turnout, and survey research suggests a relatively low level of organizational engagement among residents of Alberta as well. According to recent polling efforts, furthermore, it is not clear that, even were Alberta the site of a particularly strong civil society, such strengths would be applied to mobilization against the tar sands. A recent poll of Albertans conducted by Probe Research in 2006, which included a random sample of 500 adult residents, found that 41% of respondents held somewhat positive and 37% very positive views about development of Alberta's oil sands.[3] The most commonly cited reasons for support related to the perceived economic benefits. Other lines of questioning suggest some sources of potential grievance, however: 63% disagreed that Alberta was receiving maximum revenue from the oil sands, 87% agreed that companies could do more to protect the environment, and 50% felt the pace of development was too fast.

On Directionality

The Athabasca region's remoteness serves as a definitive feature both of the tar sands' spatiality and its directionality. Distance to markets, combined with the physical characteristics of the resource itself, raises significant implications for

---

[3]Reported at: http://www.tarsandswatch.org/overwhelming-majority-albertans-support-pause-new-oil-sands-approvals. Accessed Dec 29 2010.

material flows. In the most immediate instance, this remoteness poses a constraint to capital, the generation of which in the form of energy products requires the import of an elaborate suite of conditions of production. This should come as no surprise: the further we travel along the spectrum of non-conventional forms of fossil fuel, the greater the labour, material, and mechanical requirements to transform the earth's raw materials into energy commodities. These inputs must be transported into a region 234 miles from the nearest metropolitan centre as the crow flies, and one which furthermore is covered for half the year in ice, the other half in bog. The locally infamous Highway 63 – the only road servicing the region from points south – is today closed regularly to accommodate the enormous flat bed rigs hauling some of the largest industrial equipment ever manufactured, including everything from 400-ton trucks, shovels, separation vessels, pumps, solvents, and the like, none of which is manufactured anywhere near the location of the resource. Many employees servicing the tar pits travel by this same highway between shifts, the local region serving as home to only a tiny fraction. The results of this severely limited land transport capacity are predictable, with the accident rate inspiring some regular travellers to post the bumper sticker: "Pray for me, I drive Highway 63" (Nikiforuk 2008). Those coming from farther away commute by plane, supporting regular direct flights into Fort McMurray from as far away as St. John's, Newfoundland.

Of course the product must be moved *out* as well, and the pace at which this is done defines another critical limitation to the generation of surplus value. Bitumen is not solely a defining feature of the regional landscape: it is in many ways quite literally attached to that place, expressing a stubborn resistance to being removed. The only feasible means of transporting (and ultimately consuming) bitumen is in the form of a liquid, which it most certainly is not in its original form. The necessity of converting it into a liquid for the purposes of export is directly associated with the level of water and energy inputs required, and the environmental disruption generated. Even the sand's contribution to global climate change cannot be dis-embedded from place – the very intensity of emissions being determined by the physical properties of the resource itself, and by its distance to markets.

The greenhouse gases emitted are easily the most globally pervasive of waste flows emanating from the tar sands. Many of the remaining impacts are evidenced at the local or regional level, however, including the production of highly polluted wastewater that is stored in open ponds indefinitely, and large-scale landscape disruption associated with open-pit mining and well drilling. Other impacts travel along particular paths forged by the biosphere's natural trajectories, including water withdrawals from a major regional watershed that serves as an important ecological corridor between the Canadian Rockies and the Arctic, and the production of several air pollutants, particularly sulphuric compounds, which follow patterns of air flow to reach particular end-points near and far.

Once the resource is finally ready for distribution, it must be transported long distances to urban centres predominantly to the South and East. While early production relied on river transport, moving the product in barges either in its original solid form or in barrels, it moves today entirely by pipeline (Fig. 1.2). The fuel produced

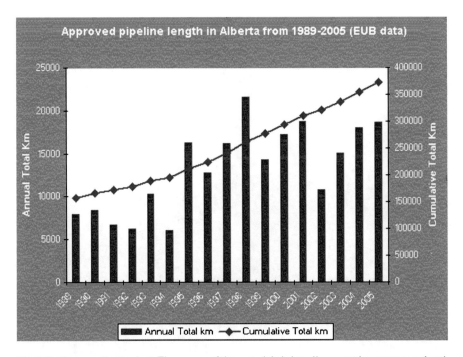

Fig. 1.2 Never-ending project. The source of the materials is http://www.environment.gov.ab.ca/. The use of these materials by the authors is done without any affiliation with or endorsement by the Government of Alberta. Reliance upon the author's use of these materials is at the risk of the end user. Accessed December 15, 2010 at: http://environment.alberta.ca/images/pipelines3.gif

from Alberta's tar deposits flows south along a continuously expanding network of pipelines that deliver 75% of the synthetic crude to US consumers; pipelines that require extraordinary capital and labour investments to construct and consequently define a relatively permanent route for delivery to urban markets. They are, furthermore, quite specialized: with solvents, unrefined and upgraded liquids all travelling along their own pipelines.

These multiple forms of spatiality and directionality, too often glossed over by scholars who exemplify the seeming ethereal qualities of global systems, are difficult to ignore in the current case, defining particular pathways of power and consequence that are relatively immutable, and in the first instances to be discussed here, intimately associated with the resource's remote spatial location. In contrast to the southerly flow of fossil fuels along the pipelines illustrated in Fig. 2.3 (see p. 34), the ancient route of the Athabasca River presented in (Fig. 1.3), from which water is extracted and into which waste leaks from the ponds, conveniently flows north, into remote regions inhabited predominantly by Aboriginal peoples – away from the powerful and towards the powerless. The Athabasca is the centrepiece of the Arctic drainage area, among the world's largest. The tar sands exemplify patterns of directionality and flow in our current era of Liquid Modernity that

Canada's continental watersheds

Fig. 1.3 The Athabasca River Watershed, centrepiece of the Arctic Drainage. Source: Canada's continental watersheds, http://www.ec.gc.ca/eau-water, Environment Canada, 2009. Reproduced with the permission of the Minister of Public Works and Government Services, 2011

highlight the extent to which we have by no means "de-materialized" – anything but, in fact – and furthermore highlight the vivid juxtaposition of material and immaterial flows, charting potentially dynamic shifts in power and consequence. Such shifts are by their nature unstable and hence unpredictable, simultaneously creating conditions for the rapid emergence of large-scale hazard and opportunities for transformation.

Situating the Tar Sands into a Global Network of Flows

The tar sands are undeniably immersed in a global network of flows. The very attention accorded to the tar sands today has arisen in large part in response to the contributions from this project to global climate change. Perspectives regarding the global significance of the absolute contribution of green house gas emissions from the Athabasca tar sands vary tremendously, with proponents attempting to minimize and opponents attempting to maximize its relativity to other sources. This numbers game is only political banter, however; the deeper significance of the tar sands' contribution to global warming is representational – as our commitment to this form of development represents a commitment to business-as-usual (namely fossil fuel-dependent) approaches to securing global energy supplies when exactly the opposite would be required to mitigate climate change. Such approaches, if pursued, represent *exacerbation* rather than mitigation of climatic change, associated as they are with an ever-increasing ratio of emissions per barrel as we necessarily dig deeper into the non-conventional fossil fuel reserve.

Within this same global system are emergent flows of capital, and capital's eternal bed partner, politics. The Athabasca is in the contradictory position of being formally under the jurisdiction of the Canadian federalist system, and yet informally subject to supra-national and global flows of markets, geo-politics, and public opinion. Governance in the twenty-first century is defined not so much by particular places or institutions but rather by flows: of wealth, commodities, environmental impacts, information, and influence. The state and corporate beneficiaries of tar sands development must thus direct their legitimation efforts towards multiple audiences, and respond to a political context that is continually in flux. The position of the tar sands in the global economy could not be more stark, most readily marked by the list of country origins of current investors, which to date includes, among others, Korea, China, Britain, Norway, Japan, and of course, the United States. Decisions regarding the export of tar sands products are also explicitly constrained by the dictates of several international trade agreements, most explicitly the North American Free Trade Agreement, which prevents Canada (or Alberta) from restricting the volume of fuels that are exported, and stipulates that the United States and Mexico have preferred buyer status (although in practice very little of Canada's energy resources ever reach Mexico).

The speed and volume of capital's circulation is matched by just one other global flow, that of information, and, particularly in the Information Age, global discourse encompasses both words and images. Indeed, images are often viewed as a more trusted source of information, emblematic of reality, and their increasing accessibility raises the possible spectre of disconnect between words and images, as when images emerge of, for example, prisoner abuse at Abu Ghraib to counter verbal insistences of military integrity. In the current instance, pontifications about Alberta's environmental stewardship are easily neutered when they cross the airwaves with satellite images of the enormous open pit mines and tailings ponds scarring the boreal landscape.

Words and images describing the tar sands have entered global circulation in the millions of gigabytes, presenting a cacophony of contested discourse on this most modern of development calamities. Only certain words and images appear to resonate, however, getting snatched up and passed on by processors functioning at multiple network nodes, while others are forever lost in the digital ether. By the same token, the theatres citizens have come to rely upon for democratic deliberation have evolved as well, both in form and audience. The discourses emanating from more traditional localized theatres, such as the political speech or public hearing, are quickly digitized for immediate global consumption, requiring disseminators to direct their message to an ever-growing audience. These forums are joined by an exponentially growing number of uber-specialized web-based discussion boards, media sources, blogs, and chat rooms.

Guided by Voices: Interpreting Future Trajectories

Travelling along these discursive channels are messages of support, critique, and resistance that transpire into ebbs and flows in legitimacy – of the tar sands as an industrial project, of the state and corporate entities that are its primary proponents, of the ideologies invoked – and by extension the potential for social transformation through and beyond this historical nexus. Evaluating this potential demands scrutiny of these relatively fixed patterns of spatiality and directionality, juxtaposed onto the current complex global system of flows.

While we are cautious about future prospects for transformation, the tar sands are unquestionably jarring to the imaginaries that have served to endorse industrialization. Those environmental hazards that often capture the imagination of risk scholars include that class of dangers that tend toward the nano-scopic: the unseen, undetected, parts-per-billion elements the very inconceivability of which strikes fear. But the gargantuan is no less undetectable, inconceivable, and thus the incomprehensibility of the scale of disruption associated with the tar sands itself has an analogous shock value. The tar sands most certainly still enjoys the support of powerful interests hailing from multiple origins, but there are some indications that the hegemony once enjoyed by these parties is now on shaky ground. The unprecedented level of discussion of the tar sands by Provincial Members of the Legislative Assembly in recent years provides one ready indicator, suggesting that while state endorsement of tar sands development did not pose a potential legitimacy threat historically, this political landscape has changed, both in terms of the source of legitimacy threat, and in terms of the universe of potentially resonant discursive frames available with which to counter that threat. Such activities are expressions of vulnerability faced by tar sands proponents, who must now rely on their waning ability to control the flow of information in an increasingly dynamic, global discursive terrain, a terrain that is distinguishable from and yet intimately connected with the local terrain. While scholars often highlight the obstacles to social movements whose task is to challenge the existing social order, observers should not discount the increasing effort required among those in power to maintain their position of dominance in Liquid Modernity, when such dominance depends far more on the delicate contours of discursive legitimacy rather than the blunter forms of coercion relied upon in days gone by.

What Follows: Outline of the Book

The next chapter provides our conceptual and methodological framework, incorporating insights from globalization theories and discourse analysis. This is followed by a chapter offering an historical perspective on discourses supporting tar sands development, alongside historical information on migration and economic patterns that have characterized Alberta's historical development. The chapter is centred on a visual history of tar sands development from the 1890s to the present, highlighting the continuities and discontinuities in these discourses, and the cultural context from which they emanate. Some recurring images of the tar sands are evident, which have organized for viewers a particular experience of the Canadian northwest and particularized views of the value of the tar sands as natural resource and as symbolic representation of Albertan identity. Tar sands images serve to instil ways of seeing that have justified state support, and represent environmental and social tradeoffs as acceptable.

Following this are three chapters describing contemporary flows of material and discourse alike. These chapters define the tar sands as an ecological irrationality, an energetic irrationality, and even an economic irrationality: irrationalities that can only be enabled in the context of disproportionalities in the distribution of wealth and suffering. These disproportionalities, furthermore, can only be politically maintained through discourse. We analyse how material flows of labour and capital, social and environmental disruptions, and energy itself are represented in discourse by corporations, politicians, social movement organizations, and citizens, and how narratives and images can obscure materiality and deflect attention from the multiple geographic scales of industrial impacts. In "Capital, Labour, and the State", we discuss briefly academic literature on Petrostates as a backdrop for subsequent presentation of information on the ownership and control of energy resources in Alberta, and the role of the energy sector in the provincial and national economies. The chapter is synthesized through the use of rent as a conceptual lens: declining organization of the material resource itself translates into increases in labour and input requirements for processing, which in turn serve as rationale for low royalties. This trend can only be expected to continue as the most accessible bitumen closest to the surface is extracted first. This chapter highlights the distortion of the regional economy due to over-investment in energy by tracing who is investing in the tar sands; what Albertans receive in the form of royalties; and forms of state subsidization (and infrastructure debt). A unique tar sands twist on the labour theory of value is highlighted (lower quality resources demand higher labour inputs). Then we take a look at rising levels of social anomie in Fort McMurray as a result of its boom town status, and the state's role in attempting to overcome capital/labour constraints to development.

In "Ecological Disruption", we come to what for many is the heart of the matter, the environmental and ecological consequences of tar sands development, and draw attention to the means by which both proponents and opponents selectively engage science and storyline to seek support for their position. We organize the chapter by type of impact, focusing on three key issues: climate change, water, and ecosystem disruption. Once again, we analyse the means by which each impact is treated discursively by proponents and opponents, with attention to how particular frames challenge or support the legitimacy of tar sands development in relation to climate change, land disturbance, and water. This is followed by "Energy Matters", in which non-conventional fossil fuel development is situated in the context of peak oil. The Energy Return on Investment associated with the tar sands is defined and discussed, including the extensive energy input requirements to support an elaborate physical infrastructure and the sustenance needs of a large labour pool in a remote, undeveloped region. Particular emphasis is placed on the inevitable *increase* in energy and resource input requirements over the course of the deposit's industrial lifetime. The ways such flows of energy are discussed, portrayed, explained away, or not mentioned at all, in narrative and imagery are analysed.

The closing section includes two chapters that ask the questions: *What can we learn* from this study about social change? And *what does the future hold* – for the tar sands, and by extension, for humanity? The final section is both a conclusion and a forecast of future plausibilities, where we ask whether or not the current conjuncture of peak oil, global warming, and global recognition of the scale of destruction in the tar sands might serve as an iconic catalyst for change, that is, as a social reckoning of our relationship to the planet. In Chap. 7, "Lessons from the Study", we integrate the material and discursive flows discussed throughout the book, depicting the historical and current trajectory of tar sands development in order to offer a systems-based explanation of events. Included is consideration of the changing faces of three fundamental political principles.

Legitimacy: We point out the key sources of legitimacy of the Alberta state's endorsement of particular development trajectories. The most significant sources of challenge to that legitimacy unveiled in this study are discussed, many of which emanate from non-historic sources beyond traditional jurisdictional boundaries.

Citizenship: We explore the current discursive contests over responsibility/ culpability, rights, and equity unveiled in this study, all from the conceptual lens of citizenship, and the ways in which our understanding of citizenship has changed in the era of global interconnectedness.

Ideology: The means by which the material and immaterial processes defining socio-ecological systems have intersected in the Athabasca tar sands in ways that challenge both our materiality and our ideologies is discussed.

In closing, we offer "A View from the Future". We sit on a precipice of both peak oil and climate change. Are we headed for catastrophe? Or will the tar sands serve as an icon? Will the tar sands say to global civil society: "Look closely: this is your future. And no one is absolved"? In this chapter we close our analysis with an exploration of observed trends leading us towards two alternative pathways, one in which our addiction to oil persists, justifying increasing dependence on non-conventional sources; the other towards reflexive transformation to post-carbon societies.

References

Alberta Chamber of Resources (2004). Oil Sands Technology Roadmap: Unlocking the Potential. http://www.acr-alberta.com/Portals/0/projects/OSTR_report.pdf. Accessed 15 January 2011.

Alberta Environment (2009a). Alberta's Oil Sands and Greenhouse Gases (GHG). http://www.environment.alberta.ca/2588.html. Accessed 15 January 2011.

Alberta Environment (2009b). Reclaiming Alberta's Oil Sands. http://www.environment.alberta.ca/2596.html. Accessed 15 January 2011.

Alberta Environment (2009c). Frequently Asked Questions – Oil Sands. http://environment.gov.ab.ca/info/faqs/faq5-oil_sands.asp. Accessed 15 January 2011.

Alberta Environment (2009d). Tailings. http://www.environment.alberta.ca/2595.html. Accessed 15 January 2011.

Bauman, Z. (2000). *Liquid Modernity*. Cambridge: Polity Press.

Beck, U. (2009). *World at Risk*. Cambridge: Polity Press.

Bentivegna, S. (2006). Rethinking politics in the world of ICTs. *European Journal of Communication 21*(3), 331–43.

Blanchard, L.A., Hinnant, C.C., & Wong, W. (1998). Market-based reforms in government: toward a social subcontract? *Administration and Society, 30*(5), 483–512.

Bunker, S.G. (1985). *Underdeveloping the Amazon: Extraction, Unequal Exchange and the Failure of the Modern State*. Chicago: University of Chicago.

Burawoy, M., Blum, J., George, S., Gille, Z. et al. (2000). *Global Ethnography: Force, Connections and Imaginations in a Postmodern World*. Los Angeles: University of California.

Canadian Association of Petroleum Producers (CAPP). (2009). Crude oil: forecasts, markets, and pipeline expansions. http://www.capp.ca/GetDoc.aspx?DocId=152951. Accessed 22 Feb 2010.

Chastko, P. (2004). *Developing Alberta's Oil Sands: From Karl Clark to Kyoto*. Calgary: University of Calgary.

Ciccantell, P.S. (2005). *Globalization and the Race for Resources*. Baltimore: Johns Hopkins University.

Ciccantell, P., Smith, D.A. (2009). Rethinking global commodity chains: integrating extraction, transport, and manufacturing. *International Journal Comparative Sociology, 50*(3–4), 361–84.

Dunbar, R.B. (2009). Existing and proposed canadian commercial oil sands projects. StrategyWest Inc. http://www.strategywest.com/downloads/StratWest_OSProjects.pdf. Accessed 15 January 2011.

Finlayson, A. (1994). *Fishing for Truth: A Sociological analysis of northern cod stock assessments from 1977 to 1990*. St John's, NL: Institute of Social and Economic Research, Memorial University.

Frank, D.J., Hironaka, A., & Schofer, E. (2000). The nation-state and the natural environment over the twentieth century. *American Sociological Review 65* (February), 96–116.

Freudenburg, W.R. & Alario, M. (2007). Weapons of mass distraction: magicianship, misdirection, and the dark side of legitimation. *Sociological Forum 22*(2),146–173.

Freudenburg, W.R., Gramling, R., & Davidson, D.J. (2008). Scientific uncertainty argumentation methods (SCAMs): science and the politics of doubt. *Sociological Inquiry 78*(1), 2–38.

Gereffi, G. & Korzeniewicz, M. (Eds.). (1994). *Commodity Chains and Global Capitalism*. Westport: Praeger.

Gillmor, D. (2005). Shifting sands. *The Walrus 2*(3), 32–41.

Habermas, J. (1975). *Legitimation Crisis*. Ypsilanti: Beacon.

Hajer, M.A. & Versteeg, W. (2005). A decade of discourse analysis of environmental politics: achievements, challenges, perspectives. *Journal of Environmental Policy & Planning 7*(3), 175–84.

Howarth, D. (2000). *Discourse*. Buckingham: Open University.

Huber, M.T. (2008). From lifeblood to addiction: oil, space, and the wage-relation in petro-capitalist USA. *Human Geography 1*(2), 42–45.

Innis, H. (1956). *The Fur Trade in Canada: An Introduction to Canadian Economic History* (Rev. ed.). Toronto: University of Toronto.

Innis, H. (1940). *The Cod Fisheries: The History of an International Economy*. Toronto: Ryerson.

Klare, M. (2008). *Rising Powers, Shrinking Planet: The New Geopolitics of Energy.* New York: Henry Holt & Company.

Lomborg, B. (2001). *The Skeptical Environmentalist: Measuring the Real State of the World.* Cambridge: Cambridge University.

Maich, S. (2005, June 13). Alberta is about to get wildly rich and powerful. What does that mean for Canada? *Macleans.* Available at: http://www.macleans.ca/article.jsp?content=20050613_1 07308_107308. Accessed 15 January 2011.

Mansell, R.L., & Schlenker, R. (2006, December). *Energy and the Alberta Economy: Past and Future Impacts and Implications. Institute for Sustainable Energy, Environment and Economy.* University of Calgary. Available at: http://www.ucalgary.ca/files/iseee/ABEnergyFutures-01. pdf . Accessed 15 January 2011.

Massey, D.B. (2005). *For Space.* London: Sage.

Massey, D.B. (1995). *Spatial Divisions of Labor: Social Structures and the Geography of Production* (2nd ed.). New York: Routledge.

Massey, D.B. (1994). *Space, Place, and Gender.* Minneapolis: University of Minnesota.

Mol, A.P.J., & Sonnenfeld, D.A. (Eds.). (2000). *Ecological Modernization Around the World: Perspectives and Critical Debates.*London: Frank Cass.

Nikiforuk, A. (2008). *Tar Sands: Dirty Oil and the Future of a Continent.* Vancouver: Greystone.

Nye, J., & Myers, J.J. (2002). *The Paradox of American Power: Why the World's Only Superpower Can't go it Alone.* New York: Oxford.

Princen, T., Maniates, M.F., & Conca, K. (2002). *Confronting Consumption.* Cambridge: MIT.

Rheingold, H. (1994). *The Virtual Community: Homesteading on the Electronic Frontier.* London: Harper-Perennial.

Shelley, T. (2005). *Oil: Politics, Poverty and the Planet.* Halifax: Fernwood.

Shove, E., & Warde, A. (1998). *Inconspicuous Consumption: the Sociology of Consumption and the Environment.* Lancaster UK: Department of Sociology, Lancaster University. http://www.lancs.ac.uk/fass/sociology/papers/shove-warde-inconspicuous-consumption.pdf. Accessed 15 January 2011.

Simon, J. (1981). *The Ultimate Resource.* Princeton: Princeton University.

Spaargaren, G. & Mol, A.P.J. (1992). Sociology, the environment and modernity: ecological modernization as a theory of social change. *Society and Natural Resources* 5:323–44.

Urry, J. (2000). *Sociology Beyond Societies: Mobilities for the Twenty-first Century.* London: Routledge.

Chapter 2
Observing Global Flows

Since the 1990s, the tar sands enterprise has evoked a collision of worldviews. At one extreme: proponents of the industry's growth and the development of what they perceive as a valuable energy source that creates investment, jobs, taxes and royalties, with reparable or justifiable costs. At the other: critics alarmed at the socio-ecological disruption associated with the extraction and consumption of 'dirty oil,' all for the sake of enriching vested interests. This chapter introduces the conceptual framework and analytical tools we use in this study to contemplate the nature, implications, and possible outcomes of this collision. Our goal is to understand better how certain courses of action with significant social and environmental consequence are justified or challenged. Such an analysis requires a close look at material realities: namely flows of money, labour, oil and waste. But even more so, it requires analysis of discursive representations and interpretations of those material realities, and the complex global system through which information, ideas, materials and political power flow. Language and imagery embody the power of material consequence, and therefore this study is first and foremost an analysis of discourse: the examination of meaning-making (Bennett et al. 2005). In this book, we are interested in the discursive strategies used by politicians, corporate representatives and their critics in open debates, in carefully crafted political speeches or reports, and in web-based communications, to frame in words and images their behaviours and concerns, to portray imagined new actions, and preferred alternative futures (See Fairclough 2003; Fairclough et al. 2003 [cited in Jacobs 2006] and Jacobs 2006). Using NVivo qualitative data analysis software as well as a archival research methods, we have collected, coded, analyzed and classified over two decades of public documents pertaining to tar sands development, including public hearing transcripts—particularly from the Oil Sands Consultations held in 2006–7; transcripts of sessions of the Alberta Legislative Assembly provided by the publicly-available *Alberta Hansard* database; as well as corporate-sponsored publications and documents; public speeches; impact assessment reports; newspaper articles and editorials; and organization websites.

D.J. Davidson and M. Gismondi, *Challenging Legitimacy at the Precipice of Energy Calamity*, DOI 10.1007/978-1-4614-0287-9_2, © Springer Science+Business Media, LLC 2011

Metaphor-Making

The primacy of discourse emanates from the simple fact that "one cannot think without metaphors" (Sontag 1991:91). To ascribe power to a phrase or an image, however, would be erroneous. The very concept of power, as the ability to do, implies agency, and only people have agency. Metaphors do not materialize out of thin air, but are socially constructed with intention. This fact empowers the metaphor-makers most of all – those agents who are successful at instilling new symbols, icons, and master frames into discourse, but also the metaphor wielders, who invoke them, often with the use of creative frame alignment practices, for the purpose of eliciting support for a particular point of view. The power to make, or wield, a metaphor with the desired effect, however, is not constructed solely through the act of discursive engagement; the agent must first be positioned so that their efforts are disseminated through social networks. This power is derived from many sources, such as the legitimacy or authority embodied in a speaker's social position (politician; scientist; specialist; layperson), or the socio-historical context of the communication event, such as a speech. Sometimes the words used by a speaker have rhetorical and symbolic resonance with social and cultural values and norms, or rekindle older traditions and sentiments in contemporary debates. Jacobs and Sobieraj (2007:2), in their study of coal field politics and Congressional debates in the United States, argue that "power ... is not simply in the discourse, but in the performance of a conflict, in the particular way in which actors mobilize discourses and reconnect the previously unconnected". All who are engaged in environmental politics go to great lengths to persuade, rationalize, or legitimate, by means of several tactics including the selective use of language and imagery, deploying metaphors, symbols and narratives (Fairclough 2003; Richardson et al. 1993; Van Dijk 1997). As social theorists, we are interested in how keywords and phrases, buzzwords and labels, or framing and rationalizations about tar sands development can direct or influence people's attitudes, thoughts and actions. The focus, however, is at a level greater than the word elements themselves. Our focus is on the powerful consequences of the dominance of particular frames and narratives and how these limit or envelop discussion, direct interpretations down certain pathways, erase major differences of fact and opinion, and win consent. Michel Foucault, well known for his analysis of the social construction of criminality, madness, and sexuality, described how analysts should approach the study of discourses (Foucault 1978:100):

> We must not imagine a world of discourse divided between accepted discourses and excluded discourse, or between the dominant discourse and the dominated one; but as a multiplicity of discursive elements that can come into play in various strategies. It is this distribution that we must reconstruct, with the things said and those concealed, the enunciations required and those forbidden,...; with the variants and different effects-according to who is speaking, his (sic) position of power, the institutional context in which he happens to be situated...; and with the shifts and reutilizations of identical formulas for contrary objectives....

In other words, discourse does not exert control by prohibiting who can speak and repressing what can be said – power is exerted by guiding and shaping how social issues are defined and publically discussed and by making key understandings appear normal or broadly accepted. The social theorist Stuart Hall explains: "a discourse provides a language for talking about ... i.e. a way of representing ... a particular kind of knowledge about a topic" ... "[and] makes it possible to construct a topic in a certain way. It also limits other ways in which a topic can be constructed" (Hall 2002:60). Some discourse analysts direct attention beyond the individual speaker to the power of privileged speech to create the "terms of reference of discussions" in ways that shape and limit complex problem-solving, such as the organization of knowledge in medicine and science, law and engineering, or public policy. Power is "materialized", as Foucault says, into legal frameworks and rights, or professional regulations and language conventions that put power in the hands of experts and members of professional organizations. In his study of the history of the James Bay Cree and the alteration of their traditional lands by hydro-electric dams, Carlson argues that legal and political processes that created a 'watershed of words' that challenged Cree understandings of the land as much as the physical alterations (Carlson 2004). Similarly, Lazuka (2006) and Ferrari (2007) both show how support for the War on Terror initiated by the U.S. state following September 11, 2001 was constructed in discourse, either by emphasizing positive portrayals of authorities (Lazuka 2006), or by adhering to "fear" and "conflict" metaphors (Ferrari 2007; Every and Augoustinos 2007).

The Tools of the Trade

The effectiveness and power of language in political persuasion can be explained by analysing the use of several discursive tools, many of which we uncover in the current analysis. In general, we draw on discourse analysis to examine the ability of critics and proponents to "affect interpretations of reality among audiences" (Benford 1997:410; see also Benford 1993; Snow et al. 1986).

We explore how verbal frames operate as discrete packages, or "schemata of interpretation" (Goffman 1974:21), to impart meaning onto particular events or processes by enclosing or limiting understandings. Framing in our work includes the ways in which politicians and corporate representatives and critics each seek to not only interpret tar sands issues, but also "define causes and solutions to a problem" (Daub 2010:119). Storylines or narratives, on the other hand, are discursive packages that include a plot, set of characters, and a set of devices that move the characters through the plot (Jacobs and Sobieraj 2007; see also Hajer 1995; Mollé 2007). In environmental controversies, Maarten Hajer argues that storylines form "a generative sort of narrative that allows actors to draw upon various discursive categories to give meaning to specific physical or social phenomena. The key function of storylines is that they suggest unity in the bewildering variety of separate discursive component parts of a problem" (1995:56).

> The 'naturalness' of narratives, or storylines, seemingly anchored in common sense, makes them very resilient.... Their appeal flows from the legitimacy they can afford policies and development programs by helping rationalize them in terms of both their intended targets and the means to be deployed to achieve those targets. By nature, they simplify and offer a

> stable vision and interpretation of reality and are able to rally diverse people around particular
> storylines. The combined actions of these people in the promotion of a storyline tend to
> coalesce into loose networks and what Hajer (1995) defined as discourse coalitions: a set of
> storylines and the actors who promote these storylines and the practices that they highlight.
> When a set of actors tries to establish hegemony and to pre-empt debate, several coalitions
> may emerge, united by their respective storylines. (Mollé 2007:7)

In other words, narratives are self-validating, producing evidence rather than the other way around, becoming "central elements in policy making" (Jacobs and Sobieraj 2007:2). Storylines deploy notions like nirvana concepts, boundary codes and icons to act as persuasive "devices that cloak policies with the symbol and trappings of political legitimacy" (Shore and Wright 1997, cited in Mollé 2007). Some narrative techniques can simplify complex events and the characters involved, increasing the legitimacy or "jurisdictional authority" of one side and delegitimizing competitor discourses (Luhmann 1989). Environmentalists become "false heroes" because saving the environment will kill the economy, and politicians become the real heroes as they create ways to preserve nature without threatening economic growth. Such tactics are certainly exercised effectively by Alberta's ruling politicians, who generate a sense of unity of purpose with the Alberta Advantage[1] and narrate themselves into a caretaker role, protecting communities in need of jobs and economic growth from the threat of downturn posed by environmental groups, concerned citizens, and other politicians, who ask for royalty increases, stronger environmental regulation, or a slowdown on project approvals. On the national and international stage, Alberta politicians become local heroes protecting the province and its energy treasure trove from federal politicians and international protocols on climate change.

Finally, another important component of our analysis is attention to implicit meanings or what is assumed, missing or unsaid in a speech or statement. As noted by Norman Fairclough, "what is 'said' in a text always rests upon 'unsaid assumptions,' so part of the analysis of texts is trying to identify what is assumed", including what is taken as given; what is assumed to be happening; and what is desirable or undesirable (2003:11, 57). We examine discourses about the future throughout our work.

Discourse and Environmental States

Often employed by social constructivists, discourse analysis, ironically, is particularly useful for eliciting the fallacy of maintaining the conceptual independence of the social and natural realms. Discursive studies of environmental politics highlight our tremendous social capacity to *interpret* the world according to personalized mental models of reality and to use those mental models to legitimate certain forms of society-environment engagement, with explicit material consequences. As has been described by Greider and Garkovich (1994), we all impose particular meanings

---

[1] "Alberta Advantage" was the Provincial slogan from 1994 until 2009. The new provincial slogan is "Freedom to create, spirit to achieve."

onto landscapes based on our experiential relationship with that environment, and also our own predilections toward particular values, such as altruism and utilitarianism. But the very ambiguity of our indicators of environmental well-being also means that, in many cases, the *only* information from which we make "observations" is discursive and this enables multiple constructions to persist. The level of danger posed by a particular toxin, for example, not to mention the very existence of that toxin, is for the vast majority of us surmised by the messages we receive through discourse, as we lack the sensory ability to detect toxins and their effects directly. The indications of the vitality of an ecosystem are equally obscure. Subsequent "claims making" activities by diverse actors attempting to influence the definition of eco-political issues (Hannigan 2006), and what counts as "fact" (Neufeld 2004), result in particular definitions and courses of action that prevail over others.

While he did not focus explicitly on environmental crises, Habermas' (1975) work on legitimacy crisis in contemporary capitalist societies is poignant here and highlights the primacy of discourse. He postulates that late capitalist societies are continuously at risk of crisis, particularly given that states are loathe to mediate many forms of *systemic* crisis (of which environmental and natural resource calamity could be included) because doing so only increases the likelihood for *legitimacy* crisis for the state. If his depiction is accurate, then those state institutions enjoying the highest levels of legitimacy may be those which are most successful at *concealing* those systemic crises through an active discursive strategy, albeit one that must necessarily undergo continuous revision and reinforcement as conditions evolve (Shenhav 2005). Analyses of the appropriateness of state roles toward the environment and resources, and the contexts and means through which common understandings are attained, contested and transformed, are particularly informative (Eckersley 2004).

Because natural resources such as timber and energy play such a fundamental role in economic development, nation-states have historically assumed jurisdiction over their management. Consequently, resource development invariably requires explicit state facilitation; simultaneously, states are liable for subsequent disruptions to air, land, and water systems (Bunker and Ciccantell 2005). Such forms of environmental degradation are associated with human health risks that tend to be inequitably distributed (Mohai and Saha 2006), posing potentially significant breaches of the social contract between citizen and state. There are certainly sound conceptual reasons to contemplate conditions under which states assume substantive responsibility for environmental well-being (cf. Frickel and Davidson 2004). But, because doing so has implications for another key source of state strength and legitimacy, namely business confidence (Block 1977) – the very conundrum described by Habermas, more often than not states have a structural incentive to avoid, rather than address, environmental problems.

Fortunately for state agents facing such a bind, the direct *indications* of environmental degradation are often obscure and are dependent upon scientific interpretation, creating the conditions for what Rothstein (2007) calls the Professional Model of the welfare state, in which state professionals invoke specialized knowledge in ways that appear impartial, privileging expert discourses while marginalizing others (Richardson et al. 1993). These circumstances contribute to an extraordinary degree of interpretive flexibility, such that interpretations themselves can determine political outcomes (Freudenburg et al. 2008; Finlayson 1994). Simultaneously, even the paradigms we embrace to depict society's relationship with nature, such as "sustainability" and "ecomodernism", are notoriously ambiguous (Hajer 1995). As such, environmental politics is particularly vulnerable to discursive manipulation, even in comparison to other political realms (Hajer and Versteeg 2005), enabling states to maintain legitimacy not by establishing sound environmental policies, but rather by invoking carefully crafted discourse that employs resonant frames, such as ecomodernism, with the potential to generate consensus and divert attention away from measurable indicators of environmental well-being (Davidson and MacKendrick 2004).

Hajer's (1995) work was among the first discursive treatments of environmental politics and remains foundational. He shows how the legitimacy of state environmental action or inaction is premised on the extent to which environmental narratives resonate with, rather than challenge, prevailing political-economic and ideological structures. In the same year, Hannigan (1995, 2006) described how selective mechanisms embedded in discourse help to determine which environmental problems get political attention and which do not. In many cases, those that do get attention are those that have been discursively stripped of their complexity. As stated by Smith and Kern (2007:5), "storylines are powerful devices through which actors make sense of complex issues without recourse to comprehensive and cumbersome explanations". But as a consequence, the complexity of eco-political issues is reduced to simplified answers. For example, "ecological modernisation is rarely considered in its full complexity. Emblems such as 'climate change' or 'clean production' or 'resource efficiency' become more amenable proxies for understanding" (2007:6). One indirect outcome of such simplification is the reinforcement of confidence in our capacity to solve environmental problems.

As we will show in various parts of the book, corporate and state supporters of the Athabasca tar sands have became adept at developing encapsulated storylines that serve to compartmentalize and simplify, concealing complexity and highlighting simplified solutions. Rather than talk of cumulative effects, tipping points, and ecological restoration, we hear about intensity targets, carbon sequestration and stakeholder partnerships. With such technological ready-fixes on hand, the flow of discourse is directed away from contradictions that might warrant deep political and social changes. Smith and Kern (2007: 6) find similar trends in the emergence of one kind of 'Transitions' discourse:

> ...the institutionalisation of this (transitions) discourse into policy has heightened rather than closed debate. This reveals the flexibility of the original storyline, and how this permits influence whilst simultaneously making it susceptible to capture. Critics argue the radical edge of the discourse has been blunted: structural goals, they argue, have been eclipsed (once again) by technocratic reforms.

Bill Freudenburg's (2006, 2005) double diversion framework neatly encapsulates much of the output of this collection of research. For Freudenburg, environmental degradation can persist due to a rather masterful "double diversion". This entails, first, diversion of *access* to resources and waste sinks into the hands of a privileged few, combined with, second, the diversion of *attention* in discourse away from the resulting disproportionalities, allowing for the perpetuation of certain privileged accounts – including in particular the unstated assumptions Fairclough highlights. One such privileged account emphasizes the "non-problematicity" of environmental degradation, as opposed to its problematicity (Freudenburg 2006:19), by framing environmental and ecological impacts as serving the greater good, while deflecting attention away from the disproportionate distribution of the "goods" in question, as well as the "bads" that result. Textual and visual information can channel public understandings, often in ways that are not immediately perceptible, and by means of discursive activities that are exercised in plain view, not behind closed doors, a form of showmanship likened to magicianship (Alario and Freudenburg 2006; Freudenburg and Alario 2007). Certain meanings become taken for granted, assumptions unquestioned, and sometimes these ways of seeing justify or hide privilege. Phrases like "environmental impact is necessary if we want jobs" can obscure the extent to which we do not all necessarily share equally in the economic benefits of environmentally costly industrial development, nor is the level of environmental impact currently associated with industrial development "necessary" to those activities. His concepts of privileged access, privileged accounts, and the diversionary framing that serves to perpetuate them are helpful for analysing how a text works when inserted into a political debate on the tar sands.

It's All in the (Power to) Name

> I want to remind the member that it's called oil sands sweet blend, not tar sands.
>
> > Honourable Guy Boutilier, Minister of Environment,
> > Government of Alberta, Legislative Assembly,
> > November 20, 2007

The bitumen deposits of northern Alberta have variously been known as tar sands or oil sands since their discovery in the 1800s. The relative accuracy of each descriptive name has been debated over the years, but in the main, "tar sands" has dominated the lay culture or vernacular, and the phrase "oil sands" has been commonly used by geologists and the business and commercial sectors, although it could be said that the geographic space, and the resource itself, "answered to both names" over the last century (Fitzsimmons 1953; Hunter 1955; Pratt 1976).

The issue of naming the resource has taken on new meaning in the last few decades in the verbal jousts between critics, who lean towards tar sands, and industry and its supporters who increasingly promote use of the term oil sands. The pattern is clear in the Alberta Legislature, for example. Figure 2.1 shows the incidence and ratio of each term as recorded in the *Alberta Hansard*, the publicly available transcript of meetings of the Alberta Legislative Assembly. From 2001 onwards, the ratio of the use of the term "oil sands" begins to increase in the official government discourse, an adoption reinforced through overt sanctions as in the opening quote from the Alberta Minister of Environment, who suggests that the issue is closed, and oil sands is the officially sanctioned term. This naming of non-conventional

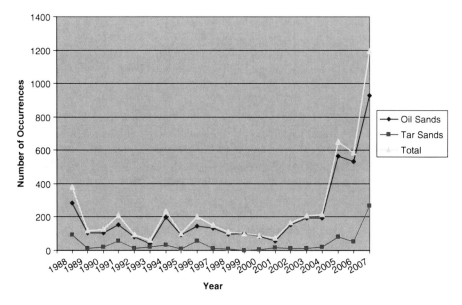

Fig. 2.1 Keyword hits over time in *Alberta Hansard*, 1988–2007. Copyright The Authors

petroleum is exemplary of the way discourses are deployed in a complex political debate. Embedded in naming conventions, symbolic shifts can invoke powerful cultural imagery, while simultaneously concealing the inconvenient and threatening, and transforming the unpalatable or controversial, tar, into the mundane – in this case, sweet blends. The power to (re)name thus embodies the power to reconstruct: to generate a sense of unity of purpose and identity and to maintain consent for processes that might otherwise be called into question. Oil – the symbolic embodiment of the industrial age, of power and modernity – portrays Alberta in a far more positive light, than tar.

This study emphasizes the value of attention to discursive processes in explaining the means by which state institutions defend their legitimacy, despite engagement in decisions and activities that could easily be interpreted as clear breaches of social contracts. Nowhere is this more blatant than in the official adoption of the term "oil sands". Neither of the terms tar nor oil are technically accurate; the resource under northeast Alberta is what geologists refer to as bitumen. The increased use of the phrase oil sands is indicative of the conscious commitment by government to brand the industry in a more positive light in the face of its critics. And yet at the same time, the attention accorded to a name exemplifies the level of effort undertaken to assert control over a potential "runaway" discourse and is suggestive of the potential tenuousness of this state's legitimacy. The discursive fields at the disposal of state actors, which offer resonance or resistance to particular framing devices, can dissipate just as readily as they emerge, particularly with the introduction of new information, new actors, and new frames. Whether the Province of Alberta will continue to be adept at manoeuvring this changing discursive landscape, only time will tell.

Visual Discourse

Visuals work alongside verbal and textual discourses as "acts of conceptual construal" (Castree and Braun 2006:167). Images and photographs shape cultural representations and perceptions and, in turn, construct the ways of seeing nature taken up by publics. They can play a particularly poignant communicative role on the international stage, where language can pose a barrier. Even maps are "not impartial reference objects, but rather instruments of communication, persuasion, and power" (Wood and Fels 1992:250). Archivist and historical geographer Joan Schwartz argues that for the last 150 years the role of photography in particular has been to "picture landscape, invest it with meaning, and articulate (our) relationship to it" (2007:966).

Remote places like the Athabasca tar sands are mediated spaces, seldom visited by outsiders before the early 1900s. Today, although the number of visitors has increased several-fold in the past century, the number of individuals who have ventured up to northeast Alberta remains relatively small, particularly in juxtaposition to the number of people who have expressed views of the Athabasca tar sands enterprise. The tailings ponds and mining sites north of the city are hidden from sight even for visitors to Fort McMurray, although some may glimpse that landscape in a company-managed field trip. The industrial zones just north of Fort McMurray have quite literally become a virtual reality for all but a select few; it has become known to citizens in Canada and elsewhere through visual representations, disseminated via digital communication networks. This remoteness also means that the images presented to global publics are derived from relatively few sources – that small number of individuals who have had access to the industrial sites themselves, and those institutions that control the collection (and dissemination) of aerial and satellite imagery. Early photographs came from an even smaller proportion of individuals, reproduced in newspapers and travellers' diaries. As interest in the resource waxed and waned from 1900 onwards, many images appear episodically in government studies and commercial reports and can be found in various public archives. Since the 1960s, the number of corporate-sponsored images has blossomed, part of the industrial record and less available to the public, or reproduced in media savvy prospectuses, press releases, and government promotions or public cultural exhibits like the Oil Sands Discovery Centre (http://www.oilsandsdiscovery.com/). Most recently, this corporate-sponsored discourse has been met with a growing onslaught of images produced by sources of resistance, all of which have become increasingly available globally through websites (corporate, government and public) which provide new sources of tar sands images. Each communication technology since 1900 has extended the reach of these representations to larger audiences.

Schwartz urges analysts to read such images to unearth their many meanings:

—not as a photograph reflecting the immanent genius of the photographer or the aesthetic qualities of the image, but as a landscape, as a visual representation of place, as a medium of geographical engagement with unfamiliar terrain. …Its content offers visual facts about the nature of land; however, the meanings invested in and generated by those facts are constructed, negotiated, and contingent—inextricably tied to the technological, historical, functional, and documentary circumstances and to the social, cultural, political, and economic contexts in which it was created, circulated, and viewed. (2003, pp.109-10)

For Schwartz, nature photographs have contributed "to the construction of an imaginative geography of Canada" and been used over time by various interested parties "not simply to document reality, but to evoke it" (Schwartz 2003: 977). In other words, images have an unrecognized "epistemic quality" which can constrain people's thinking (Hajer and Versteeg 2005) and photographic representations of material nature often exude, as anthropologist Clifford Geertz once wrote of religion, "an aura of factuality ... that seem(s) uniquely realistic" (Geertz 1966, cited in Vitalis 2009:IX).

We use a number of visuals throughout the book to illustrate how images of the tar sands express meanings about the resource and construct certain "factual" representations of the biophysical setting, or the practices of tar sands extraction and upgrading. We compare and contrast the contemporary period with other historical conjunctures where state legitimacy, geopolitics of oil security, and tar sands industrial growth have intersected. We specifically select those images that have been used by certain political interests to portray a particular narrative regarding the tar sands, used in print media, web sites, and internet exhibits (Dodge and Perkins 2009). We analyse what Hall (2002) calls "the dominant or preferred readings" of these images and their contribution to meaning-making in tar sands politics, especially the way some images have the power to politicize, and others to *de*politicize certain peoples, places and ecologies. In the next chapter, we concentrate on the photographs taken by the site's first European visitors on government-sanctioned visits, during the century beginning in 1880, offering a pictorial history of the early stages of tar sands exploration and field science and the initial commercial stage in the 1940s, to the first "boom" in the 1970s. Some of the historic photos used are presented in the Oil Sands Discovery Centre exhibits, in Fort McMurray.[2] In later chapters, we introduce contemporary photographs taken by a wider variety of sources, including the aerial photographs, satellite images and clandestine photographs taken by critics to elicit global protest against the scale of mining and pollution occurring in the Fort McMurray region, as well as corporate and state promotional media, developed in part with the specific intention of countering the growing opposition movement. Changes in photographic technology, from early stills, to aerial, digital and satellite photography, can in and of themselves alter our gaze, and we follow those changes in our analysis. But, the relationship between photographic images, ideology, and legitimation, how photographic representations shape collective views of the industrial use of nature, remain central to our reading. One such photograph has become an iconic symbol of Alberta's energy heritage, the spudding of the Leduc Number 1 oil well on 13 February 1947 (Fig. 2.2).

The first large scale oil deposit discovery in Alberta, Leduc No. 1 has come to personify a set values that have been ascribed to the oil industry in Alberta. This photograph has been used repeatedly in the popular media to represent not solely

---

[2]On the social construction of museum exhibits see Dirks et al. (1994); Boswell and Evans (1999).

Fig. 2.2 http://en.wikipedia.
org/wiki/Leduc_No._1.
GNU Free Documentation
License

Alberta's entry into the modern oil industry, but also a metaphor for Alberta's prosperity, our identity as "oil" country. Hariman and Luciates (2007) argue that iconic images represent unspoken civic virtues associated with an historic event. In the Leduc No. 1 image, the dominant reading communicates a message. Devoid of people, it nevertheless makes a strong statement about the role of humans in nature and the character of the men (there were few women) whose ingenuity and hard work have harnessed it. It states in no uncertain terms that human industry can control or bend nature to our economic purposes. And that the business risks and physical perils of oil extraction – taken by those who work in and run the oil industry – are worth the economic gamble (Thompson 1998; Friedel 2008).

But does this photograph simply capture an historic moment? In actuality, this apparently spontaneous depiction of the discovery of oil at Leduc was carefully stage-managed over 2 days – the gas flare-off ignited "on command" for full media effect – by Imperial Oil's manager, for the benefit of reporters and government officials (Thompson 1998:150). A contrivance, the image does not simply reflect, but *creates* a link between entrepreneurship and resource exploitation as inter-related virtues. In turn, it establishes a singular, utilitarian vision of nature that becomes unquestioned, reducing social critique and meaningful debate about trade-offs.

In Canada, images of natural resource exploitation – mines, logging trucks, oil rigs, dams, and fishing boats – are routinely depicted as part of the industrial and cultural aesthetic of the nation. Such images offer representations of certain elements of nature (and society) as expendable – especially as governments and corporations unleash new sources of energy and wealth. Most people draw on preferred readings of resource extraction without reflection: "they do so because these cultural structures – by virtue of their institutionalization – are the most widely available, retrievable, and familiar" (Gabara 2006). In northern Alberta's tar sands, nature's offerings became photographic clichés that recur across the visual record – the treasure troves of tar and oil weeping down the riverbanks; prestigious visitors with cupped-hands full of rich black sand; and humans dwarfed by geological deposits and later by mining equipment. Schwartz' socio-historical approach directs our attention to the pre-texts of viewing, that is, the "intellectual baggage which needs to be teased out by thinking not only *about*, but also *with*, the photograph, by using it not simply as a source of facts, but as a mode of inquiry, by bringing context to the photograph, better to understand the context. What, then, did [in our case photographic images] mean to those who read the texts, saw the images, studied the maps, and examined the specimens…?" (Schwartz 2003:114).

Mapping Flows: Words and Images in the Network Society

Changes in communications technology have enabled not only enhanced ability to alter images, and to expand the scale of that imagery, but have also enabled rapid dissemination through a global communications network. These changes represent just one means by which context matters. While the rhetorical tactics embodied in discourse may have withstood the tests of time, the role of discourse in politics today is qualitatively different than was the case when the Athabasca tar sands were first discovered a century ago. To a much greater extent than has ever been the case in history, political engagement today can be premised solely on the basis of discourse, rather than personal experience or nation-state membership. Advances in communications technology have enabled a relinquishment of the reins of communication channels by established state, corporate and media channels, although vested interests most certainly have sustained a dominant position. This global communications network is superimposed upon much older, but nonetheless evolving, networks of capital and resource flows.

As noted by several globalization theorists (e.g. Mann 1993; Urry 2000; Giddens 1999; Bauman 2000), contemporary societies are just plain messy; messier in many cases than the conceptual mainstays of sociology – nation-state, class, etc. – allow. There have been some recent theoretical innovations, however, that offer researchers new resources. Certainly, Manuel Castells' work on the Network Society (2010) – a term originally coined by van Dijk (1991) – is notable. In this work, Castells explores how social structures and processes increasingly become organized around electronically processed information networks, networks which can be described by multiple linkages and the hubs where they intersect. Urry (2000) offers a conceptual framework, consisting of "scapes" and "flows", which is especially useful for our analysis. According to Urry, scapes – including communication networks – represent the canvas upon which social processes unfold: "the networks of machines, technologies, organizations, texts and actors that constitute various interconnected nodes along which the flows can be relayed.... Once particular scapes have been established, the individuals and especially corporations within each society will normally try to become connected to them through being constituted as nodes within that particular network" (Urry 2000:35). These scapes are not purely ethereal, however, but rather:

> constitute relatively determinate networks or chains of exchange within a space. The world of commodities would have no 'reality' without such moorings or points of insertion, or without their existing as an ensemble ... of stores, warehouses, ships, trains and trucks and the routes used .... Upon this basis are superimposed—in ways that transform, supplant or even threaten to destroy it—successive stratified and tangled networks which, though material in form, nevertheless have an existence beyond their materiality: paths, roads, railways, telephone links, and so on. (Lefebvre 1991: 402–3)

Because such scapes constitute a physical and relatively fixed presence, the channels constructed also describe patterns of equity, as they empower some, and pass by others (Graham and Marvin 1996). Even the seemingly ubiquitous networks, such as that of digital information, are constrained by the direction of investments into microwave towers, fibre-optic cables, and the like, which must go somewhere in space, and thus favour some routes over others. This is not to say that the maps describing the global network society are static, but they most certainly do not evolve instantaneously. Consider, for instance, the construction of an oil pipeline, which by virtue of the sheer magnitude of investment required for its construction, "fixes" a channel of a given volume, between a particular producer and a particular consumer (or processor, or distribution point), for several decades (Fig. 2.3). Others are more malleable, such as the shifts in flight patterns – Canadian airlines were relatively quick to respond to patterns of labour supply and demand with direct flights between Newfoundland and Fort McMurray, for example. This map of *intentional* networks – those constructed for desired effects – also has unintended consequences, however, including the obvious, like closing doors to alternative pathways, and the not-so-obvious, like the hollowing out of communities in Newfoundland of much-needed human capital. The scapes defining contemporary society also include *un*intentional pathways for the flows of risk, waste, and so on. These channels are carved by biospheric processes such as water cycles, air sheds, river basins and ocean currents.

The flows in question "consist of peoples, images, information, money and waste, that move within and especially across national borders and which individual societies are often unable or unwilling to control directly or indirectly" (Urry 2000: 36). Each of these flows must necessarily move at different paces, however, paces that are dictated to a great extent by the physical properties of the elements themselves. The flow of energy sources is an elemental factor of our economies. One of the most ubiquitous commodities on the planet, coal, is an extremely heavy solid that can withstand a variety of environmental conditions, but that has a relatively

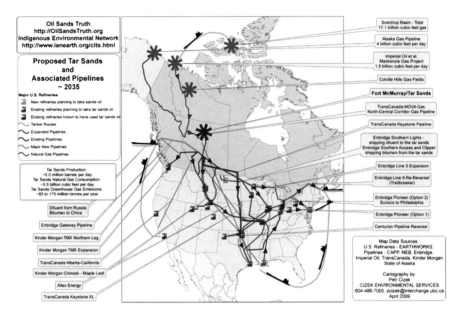

Fig. 2.3 http://oilsandstruth.org/updated-continental-maps-pipelines-2035. Used with permission

low value-to-volume ratio, and thus it makes economic sense to minimize the distance travelled and makes the containers as large as possible for the sake of efficiency. Natural gas, on the other hand, is literally lighter than air, but can only travel by pipeline without the introduction of rather tedious, expensive, and extremely dangerous liquefaction processes and has a nasty tendency to explode. Many renewable forms of power, such as wind and solar, can only be transported directly by power line, along which the strength of that current dissipates, and thus they have an extremely limited geographical market reach. Digital forms of information and money can move virtually instantaneously by comparison. Humans, of course, have their own sets of requirements while in transit.

What is of particular interest in this study are the intersections among flows in space and time, particularly the intersections between environmental and resource flows, and the discourses employed to describe them. Discourse does not exist in a vacuum, after all. History is defined not just by ideas, but rather by their inherent materiality; by the ever-evolving inter-relationships between the material world on the one hand, and on the other, how we talk about that world and our place in it. The evaluation of discursive legitimation in environmental and natural resource politics is relatively new, and yet holds tremendous intellectual merit, given the dynamic intersections between material and non-material, objective and subjective matters that both highlight the fluidity of modernity and ground that modernity in physical space and historical time (Urry 2010; Freudenburg 2006). As noted by Escobar (1996), a materialist analysis *must* encompass a discursive analysis, as our material "reality" cannot be separated from our discursive representations of it. The very expansion of capital onto nature presupposes a discursive consumption of nature *as* capital (O'Connor 1988, 1998).

The effects on society and individuals of such complexity have been remarked upon by several scholars, the most notable of which are the escalating pace and tandem decline in predictability of social change, or what Archer (1995) calls processes of morphogenesis. We highlight a few more specific effects of particular interest to the current analysis. The first such effect of interest is social equity. The multiple intended and unintended consequences of global flows which favour some and compromise others have engendered the emergence of "resistance identities" (Castells 1997: 356), bound not by geography but by constructed identities defined by their relationship to these flows. In turn, a significant feature of contemporary social movement activity is that information technologies have enabled resistance movements some success (Evans 2000). Indeed, the anti-globalization movement is one of the most compelling trends in global politics today (Buttel and Gould 2004).

We should not wax too optimistic about these trends, however, without critical consideration of some not-so-new but rather enduring forms of inequity. The power imbalance between staples providers and advanced economies that consume them describes an age-old dependency relationship that has deepened as the scarcity of critical resources has expanded the geographical confines of their exploitation by states and corporations. In addition, prospects for the formation of international resistance identities are defined not solely by the geographic map of the communication netscape described above, but also by the individual-level human capital needed to access that network. While the rise of digital media has undoubtedly opened political doors to members of the middle classes, it has provided an additional form of marginalization for the world's poor.

A second effect of interest, called "nominalization" by Fairclough (2003: 11) "contributes to what is … a widespread elision of human agency in and responsibility for process in accounts of the 'new global economy' … nominalizations reduce agency and obscure who is responsible for change". The discourses swirling around the Athabasca tar sands are replete with examples of nominalization, in which decisions appear inevitable and responsibility for actions taken is lost. To take just one example, a repeated trope in tar sands discourse used by corporate and state interests goes something like this: "the global demand for oil is driving the pace and scale of oil sands expansion". To a significant degree, nominalization represents an individual response to a seemingly overwhelming social system. Increasing mobility and fluidity in our "Liquid Modernity" introduce feelings of detachment and uncertainty (Bauman 2000). Humans increasingly must manoeuvre within circumstances "which are not of their own making" (Urry 2000:14), encouraging a sense of personal disempowerment. But nominalization is not *only* personal, it has its expression in organizational behaviour as well. Our very organizational efforts to simplify complex environments, by means of specialization and compartmentalization, for example, narrow an agent's vision of the consequences of her actions. The focus of attention is reduced to that one purchase, that one licence application, rather than the cumulative effects of multiple decisions and transactions.

This book is not primarily a theoretical treatment, and thus this presentation of theory is notably brief, presented namely for the purposes of identifying those of our own theoretical musings that have influenced our empirical analysis. We spend much of our effort characterizing the scapes and flows that define tar sands politics, but our purpose is not merely descriptive. Our goal is to identify certain trends and conditions in this political theatre that may amount to collective tendencies toward differing future pathways. Will we continue along a trajectory of increasing reliance on non-conventional fossil fuels, enduring all the calamities embodied in that pathway, or will the Athabasca tar sands serve as an icon, fostering support for, and development of the means toward, transition to alternative pathways?

References

Alario, M. & Freudenburg, W.R. (2006). High-risk technology, legitimacy and science: the U.S. search for energy policy consensus. *Journal of Risk Research 9*(7):737–53.

Archer, M. (1995). *Realist Social Theory: The Morphogenetic Approach*. Cambridge: Cambridge University.

Bauman, Z. (2000). Liquid Modernity. Cambridge: Polity Press.

Benford, R.D. (1993), Frame disputes within the nuclear disarmament movement. *Social Forces 71*(3), 677–701.

Benford, R. (1997). An Insider's Critique of the Social Movement Framing Perspective. *Sociological Inquiry 67*(4),409–430.

Bennett, T., Grossberg, L. & Morris, M. (Eds.). (2005). *New Keywords: A Revised Vocabulary of Culture and Society*. Oxford: Blackwell.

Block, F.L. (1977). The ruling class does not rule. *Socialist Revolution 7*(33), 6–28.

Boswell, D. & Evans, J. (Eds.). (1999). Representing the nation: a reader. *Histories, Heritage and Museums*. London: Routledge.

Bunker, S.G. & Ciccantell, P.S. (2005). *Globalization and the Race for Resources*. Baltimore: Johns Hopkins University.

Buttel, F.H. & Gould, K.A. (2004). Global social movement(s) at the crossroads: some observations on the trajectory of the anti-corporate globalization movement. *Journal of World Systems Research 10*(1), 37–66.

Carlson, H.M. (2004). A watershed of words: litigating and negotiating nature in Eastern James Bay, 1971-75. *The Canadian Historical Review 85*(1), 63–84.

Castells, M. (2010 [1996]). *The Rise of the Network Society*. Oxford: Wiley-Blackwell.

Castells, M. (1997). *The Power of Identity*. Oxford: Blackwell.

Castree, N. & Braun, B. (2006). Constructing rural natures. In Cloke, P., Marsden, T., & Mooney, P. (Eds.). *Handbook of Rural Studies*. London: Sage.

Daub, S. (2010). Negotiating sustainability: climate change framing in the Communications, Energy and Paperworkers Union. *Symbolic Interaction 33*(1),115–140.

Davidson, D.J. & MacKendrick, N.A. (2004). All dressed up and nowhere to go: the discourse of ecological modernization in Alberta, Canada. *Canadian Review of Sociology and Anthropology 41*(1), 47–65.

Dirks, N., Eley, G. & Ortner, S. (Eds.). (1994). *Culture/Power/History: A Reader in Contemporary Social Theory*. Princeton: Princeton University.

Dodge, M. & Perkins, C. (2009). The view from nowhere? spatial politics and cultural significance of satellite photography. *Geoforum 40*(4), 497–501.

Eckersley, R. (2004). *The Green State: Rethinking Democracy and Sovereignty*. Cambridge: MIT.

Escobar, A. (1996). Construction nature: elements for a post structuralist political ecology. *Futures* *28*(4), 325–343.

Evans, P. (2000). Fighting marginalization with transnational networks: counter-hegemonic globalization. *Contemporary Sociology 29*(1), 230–241.

Every, D. & Augoustinos, M. (2007). Constructions of racism in the Australian parliamentary debates on asylum seekers. *Discourse and Society 18*(4), 411–36.

Fairclough, N. (2003). *Analysing Discourse: Textual Analysis for Social Research.* London: Routledge.

Fairclough, N., Graham, P., Lemke, J., & Wodak, R. (2003). Introduction. *Critical Discourse Studies 1*(1), 1–7.

Ferrari, F. (2007). Metaphor at work in the analysis of political discourse: investigating a 'preventive war' persuasion strategy. *Discourse and Society 18*(5), 603–25.

Finlayson, A. (1994). *Fishing for Truth: A Sociological Analysis of northern cod stock assessments from 1977 to 1990.* St. John's: Institute of Social and Economic Research, Memorial University.

Fitzsimmons, R. C. (1953). *The Truth About Alberta Tar Sands: Why Were They Kept out of Production?* Edmonton, Alberta. Pamphlet.

Foucault, M. (1978). *The History of Sexuality.* New York: Vintage.

Friedel, T. (2008). (Not so) crude text and images: staging Native in 'big oil' advertising. *Visual Studies 23*(3), 238–254.

Freudenburg, W.R. (2006). Environmental degradation, disproportionality, and the double diversion: reaching out, reaching ahead, and reaching beyond. *Rural Sociology 71*(1), 3–32.

Freudenburg, W.R. (2005). Privileged access, privileged accounts: toward a socially structured theory of resources and discourses. *Social Forces 84*(1), 89–114.

Freudenburg, W.R. & Alario, M. (2007). Weapons of mass distraction: magicianship, misdirection, and the dark side of legitimation. *Sociological Forum 22*(2), 146–173.

Freudenburg, W.R., Gramling, R., & Davidson, D.J. (2008). Scientific Uncertainty Argumentation Methods (SCAMs): science and the politics of doubt. *Sociological Inquiry 78*(1), 2–38.

Frickel, S. & Davidson, D.J. (2004). Building environmental states: legitimacy and rationalization in sustainability governance. *International Sociology 19*(1), 89–110.

Gabara, E. (2006). Recycled photographs: moving still images of Mexico City, 1950-2000. In Schwartz, M.E. & Tierney-Tello, M.B. (Eds.), *Photography and Writing in Latin America: Double Exposures.* Albuquerque: University of New Mexico.

Geertz, C. (1966). Religion as a cultural system. In Banton, M. (Ed.), *Anthropological Approaches to the Study of Religion.* New York: Praeger.

Giddens, A. (1999). *Runaway World: How Globalization is Reshaping Our Lives.* London: Profile.

Goffman, E. (1974). *Frame Analysis.* Cambridge: Harvard University.

Graham, S. & Marvin, S. (1996). *Telecommunications and the City.* London: Routledge.

Greider, T. & Garkovich, L. (1994). Landscapes: the social construction of nature and the environment. *Rural Sociology 59*(1), 1–24.

Habermas, J. (1975). *Legitimation Crisis.* Ypsilanti: Beacon.

Hajer, M.A. & Versteeg, W. (2005). A decade of discourse analysis of environmentalpolitics: achievements, challenges, perspectives. *Journal of Environmental Policy & Planning 7*(3), 175–84.

Hajer, M.A. (1995). *The Politics of Environmental Discourse: Ecological Modernization and the Policy Process.* Oxford: Clarendon.

Hall, S. (2002). The west and the rest: discourse and power. In Shech, S. & Haggis, J. (Eds.), *Development: A Cultural Studies Reader.* Oxford: Blackwell.

Hannigan, J.A. (1995). *Environmental Sociology.* London: Routledge.

Hannigan, J.A. (2006). *Environmental Sociology* (2nd ed.). London: Routledge.

Hariman, R. & Luciates, J.L. (2007). *No Caption Needed: Iconic Photographs, Public Culture, and Liberal Democracy.* Chicago: Chicago University.

Hunter, G. (1955). Athabasca tar sands. A photographic essay. *The Beaver.* Winnipeg: Hudson's Bay Company. (1955), 14–19.

Jacobs, K. (2006). Discourse analysis and its utility for urban policy research. *Urban Policy and Research 24*(1), 39–52.

Jacobs, R. & Sobieraj, S. (2007). Narrative, public policy, and political legitimacy: congressional debates about the nonprofit sector, 1894-1969. *Sociological Theory 25*(1), 1–25.

Lazuka, A. (2006). Communicative intention in George W. Bush's presidential speeches and statements from 11 September 2001 to 11 September 2003. *Discourse and Society 17*(3), 299–330.

Lefebvre, H. (1991). *The Production of Space*. Oxford: Blackwell.

Luhmann, N. (1989). *Legitimation durch Verfahren*. Frankfurt am Main: Suhrkamp.

Mann, M. (1993). *The Sources of Social Power, Vol. 2*. Cambridge: Cambridge University.

Mohai, P. & Saha, R. (2006). Reassessing racial and socioeconomic disparities in environmental justice research. *Demography 43*(2), 383–99.

Mollé, F. (2007). Sacred cows, storylines and nirvana concepts: insights from the water sector. Paper presented at World Water Week, Special Session on Water, Politics and Development. Stockholm, August.

Neufeld, M. (2004). Pitfalls of emancipation and discourses of security: reflections on Canada's 'security with a human face.' *International Relations 18*(1), 109–123.

O'Connor, J. (1998). *Natural Causes: Essays in Ecological Marxism*. New York: Guilford.

O'Connor, J. (1988). The second contradiction of capitalism. *Capitalism, Nature, Socialism* (1):11–38.

Pratt, L. (1976). *The Tar Sands: Syncrude and the Politics of Oil*. Edmonton: Hurtig.

Richardson, M., Sherman, J. & Gismondi, M. (1993). *Winning Back the Words: Confronting Experts in an Environmental Public Hearing*. Toronto: Garamond.

Rothstein, B. (2007). Creating state legitimacy: the five basic models. Annual Meeting of the American Political Science Association, Chicago Aug 28-Sept 2. http://www.law.yale.edu/documents/pdf/intellectual_life/rothstien_creating_state_legitimacy.pdf. Accessed 25 July 2011.

Schwartz, J. (2003). More than 'competent descriptions of an intractably empty landscape': a strategy for critical engagement with historical photographs. *Historical Geography 31*.105–30.

Smith, A. & Kern, F. (2007). The transitions discourse in the ecological modernisation of the Netherlands. SPRU (Science & Technology Policy Research), University of Sussex, UK. Paper for the Earth Systems Governance conference in Amsterdam, 24-26 May. http://www.2007amsterdamconference.org/Downloads/AC2007_SmithKern.pdf. Accessed 25 July 2011.

Snow, D.A., Rochford, E.B. Jr., Worden, S.K., & Benford, R.D. (1986). Frame alignment processes, micromobilization, and movement participation. *American Sociological Review 51*(4): 464–481.

Shenhav, S.R. (2005). Concise narratives: A structural analysis of political discourse. *Discourse Studies 7*(3), 315–35.

Shore, C. & Wright, S. (1997). Policy: a new field of anthropology. In Shore, C. & Wright, S. (Eds.). *Anthropology of Policy: Critical Perspectives on Governance and Power*. London: Routledge.

Sontag, S. (1991). *Aids and its Metaphors*. Harmondsworth: Penguin.

Thompson, J.H. (1998). *Forging the Prairie West: The Illustrated History of Canada*. Toronto: Oxford University.

Urry, J. (2000). *Sociology Beyond Societies: Mobilities for the Twenty-first Century*. London: Routledge.

Urry, J. (2010). Consuming the Planet to Excess. *Theory, Culture & Society 27*(2–3),191–212.

van Dijk, T. (Ed.). (1997). *Discourse as Social Interaction*. London: Sage.

van Dijk, J. (2006 [1991]). *The Network Society: Social Aspects of New Media* (2nd ed.). London: Sage. Orig pub in Dutch: 1991. *De Netwerkmaatschappij: Sociale aspecten van nieuwe media*. Bohn Stafleu en van Loghum, Houten.

Vitalis, R. (2009). *America's Kingdom: Mythmaking on the Saudi Oil Frontier*. London: Verso.

Wood, D., Fels, J. (1992). *The Power of Maps*. New York: Guilford.

Chapter 3
Visualizing the Tar Sands Through Time

Incubation

The Athabasca tar sands may have been born virtually overnight in the international media, but only after a decades-long incubation period during which the maternal side of Albertans shone. This incubation is captured – the baby itself in fact constructed – in discourse as much as in industriousness. And within this discourse, it is the images that have withstood the test of time, marking each trimester in vivid relief. Some would argue that the use of images in contemporary politics and social protest marks a recent "visual turn" in post- or late modern society. But modern analysts often overlook long-standing visual representations, in the case of the tar sands, well over 100 years of visual storytelling, which has established an authoritative industrial discourse in support of corporate investment, government assistance, the inevitability of commercial-scale exploitation, and ultimately human domination of a passive nature. As argued by Schwartz and Ryan (2003), historic photographs embed the Canadian landscape with constructed meanings and inform the public's normative and scientific imaginings of Canada's northwest and its natural systems, and this is particularly true of the tar sands, a mundane naturally occurring seepage turned coveted national treasure.

This visual history is important to our understanding of tar sands development, because it tells the story of just how manufactured this enterprise is. Over the last century, billions of dollars of government support have been spent to see this imagined treasure to fruition. That level of state investment required politicians to justify such an extraordinary expenditure to taxpaying citizens *and* encourage the corporate elite to cough up some of their own surplus, all for an uncertain enterprise that ultimately would not see the light of day for generations. Remote even by today's standards, and visited by few outsiders before the early 1900s, the Athabasca tar sands have always been a mediated space that has become known to most of the world through the diaries, reports, photographs, and film images developed by explorers, travelers, government employees, and early industrialists. For most contemporary Canadians, the Athabasca tar sands have become synonymous with these

D.J. Davidson and M. Gismondi, *Challenging Legitimacy at the Precipice of Energy Calamity*, DOI 10.1007/978-1-4614-0287-9_3,
© Springer Science+Business Media, LLC 2011

Fig. 3.1 Library and Archives Canada. PA-038166. Geological Survey of Canada, G.B. Dowling, 1892. This image from the Geological Survey depicts two men dwarfed by a black swath of tar sands standing alongside the river bank of the Athabasca River in 1892. It suggests the enormity of the resource, its potential, and the apparent easy availability of the tar sands resource

iconic visual representations – riverbanks weeping tar; cupped hands full of oil sands; scientists boiling tar sands in the field; workers dwarfed by giant mining equipment; and industrial plants emerging on the boreal frontier (Fig. 3.1). Even today, visitors to the region seldom see the mines, tailing ponds, and processing plants; instead they glimpse the landscape through company-managed field trips, but far more often through the images of web sites and magazines, or at interpretive displays, films, and photographic displays in the Oil Sands Discovery Center and gift shop in downtown Fort McMurray.

In this chapter, we take an historical approach to examining how images have been used over time to represent and narrate both past and present processes of tar sands development. In later parts of the book, we offer a number of contemporary images that supporters and critics use to "imagine possible futures" (Fairclough 2009:322); futures that, as we shall discuss, either create acquiescence and defer-ence by publics to corporate and technological solutions to climate change and peak oil, or open up moral and political alternatives.

A Seed Is Planted

Cultural perceptions of Canada's northern landscapes often presented a complex frontier, at once characterized as barren, cold, and unyielding wilderness, yet a place where personal freedom and fortitude can flourish, with enormous economic reward. At the turn of the last century, government officials, commercial promoters, and historians constructed Canada's northwest landscape as a storehouse of hidden riches in the attic, encouraging a culture of discovery that echoes across the historic visual record. Early images emerged at a timely point in history, when Canadians were primed for such pipedreams, in the years just before and after the Klondike gold rush. Some of the first popular images of the tar sands were in fact taken by members of the gold-digging entourage that traveled the Athabasca River to the Yukon in search of their fortune in 1898. For some, the tar sands would become their Klondike. Most early travelers' gazes were confined to these river valleys, quite literally confirming Harold Innis's claim that transportation corridors played a key role in moulding the social structures that would define Canada's staples economy. In the northwest in the early 1900s, travelers along the Athabasca River north of Fort McMurray reported miles of bitumen seams and occasional free-flowing oils and tars, and such sights would excite the imaginations of journalists, governments, and others, at a time when the world was only just discovering the potential for oil.

Early Canadian government efforts to extend national sovereignty in the west through the Geological Surveys of Canada had begun in the late 1800s, establishing Canadian rule over the Athabasca territory at a time when the United States and Britain coveted the northwestern frontiers of a young Canadian nation. Surveyor and engineer reports on minerals and natural features including the commercial potential of the tar sands sutured the area into "the transcontinental imaginary of Canada" (Zeller 1987). Confirming lay observations by adventurers, explorers, and naturalists, government experts in the District of Athabasca reported a landscape and topography with visible layers of tar. Their photographs often include early efforts by entrepreneurs to drill for and extract oil from tar deposits. Dowlings' photographic image (above) testified to even earlier reports in 1882 by Dr. Robert Bell, who became chief geologist and acting director of the Geological Survey of Canada. Bell was considered Canada's leading field scientist, and his claims that beneath the sandy pitch and tar bands lay "pools of petroleum," embellished with images of bitumen bursting from the river banks, stimulated the Canadian Senate in 1888 to commission a Committee Report to Parliament (Zaslow 1971:79):

> The evidence submitted to your committee points to the existence in the Athabasca and Mackenzie Valleys of the most extensive petroleum field in America, if not the world. The uses of petroleum and consequently the demand for it by all nations are increasing at such a rapid ratio, that it is probable this great petroleum field will assume an enormous value in the near future and will rank among the chief assets in the Crown domain of the Dominion.

The following pages present a selection of popular images and eyewitness accounts of the tar sands between 1900 and the 1970s, taken from archival sources, diaries and reports of journalists, scientists, surveyors, and commercial pioneers. While today there are many competing images of the tar sands industry, for much of its history one voice (and its visual conventions or tropes) shaped understandings of the industry's emergence and trials. Most notably, two key patterns cut across the visual record: the application of human ingenuity and scientific and technological expertise to release oil from the chemical bonds of its bitumen form; and corporate and government determination to overcome the physical and economic challenges of opening up the northern frontier to the new industry. Like all powerful social facts, consumed by early Canadians eager to create their own national identity, many of these well-constructed images now appear familiar to modern readers, part of our mental inventory of representations of Western Canadian history and culture.

The Messengers

One of those en route to the Klondike gold rush was Count Alfred Von Hammerstein. Like many others in this entourage, he noticed, and photographed, the tarry seams. But unlike most, his visions of gold riches readily shifted to a more pragmatic but no less potentially lucrative vision. As the subtitle "flowing asphaltum" in the Figure above suggests, Von Hammerstein saw in the tar an opportunity to help pave the roads of Alberta's new capital city of Edmonton. And he also had the means. A newspaper owner and financier of dubious reputation, Von Hammerstein found investors in western Canada and Britain and returned to drill wells along the river. *Le Courrier de L'ouest* (1910) reported on 12 January of 1910 that Von Hammerstein's investors had committed over five million dollars to his "Athabasca Oil and Asphalt Company (Figs. 3.2–3.4)."

Others brought the Athabasca tar to the public imagination through literature, including in particular the adventurer and writer Agnes Deans Cameron, who wrote:

> In all Canada there is no more interesting stretch of waterway than that upon which we are entering. An earth-movement here has created a line of fault clearly visible for seventy or eighty miles along the river-bank, out of which oil oozes at frequent intervals. ... Tar there is... in plenty.... It oozes from every fissure, and into some bituminous tar well we can poke a twenty foot pole and find no resistance (Deans Cameron 1909:71).

Crafted in a wilderness travel-journal style and yet enhanced by the authority of new communication technologies like the Underwood typewriter (1895) and the Kodak camera (1888), Deans Cameron's descriptions of the natural history of the region, supplemented by photographs, offered up a northern landscape overflowing with riches and begging to be industrialized (Roy 2005; O'Leary 2004:20). Able to witness the early efforts of industrialists like Von Hammerstein, Deans Cameron highlighted "the labor and determination which in this wilderness has erected these giant derricks" (Cameron 1909:70). The Count's endeavors to find "elephant pools of oil" were ultimately unsuccessful, but her book, *The New North*, became a best

OIL SAND EXPOSURE (Height appr. 280 feet) on claim near Fort McMurray. Millions of tons of material in sight. Note the oil seepages, indicating the richness of material.

This is a solid mass of Bituminous Sands.

Fig. 3.2 This image appears in the Alberta Provincial Archives without a source, but carries the descriptive subtitle: "OIL Sands Exposure – (height approx. 280 ft) on claim near Fort McMurray. Millions of tons of material in sight. Note the oil seepages, indicating the richness of material. This is a solid mass of Bituminous Sands." PAA 77.178.22 with permission

Fig. 3.3 The same image,
with the script "Tar Sands
and flowing Asphaltum in the
Athabasca District ca. 1908."
is found in the Alfred Von
Hammerstein collection/
Library and Archives Canada
/PA-029259

seller and made Deans Cameron a media celebrity. She would lecture across Canada
and the United States about her "Journeys through Unknown Canada." Later, as a
Canadian Government representative in Britain, with photographic images on dis-
play, she offered a popular series of public talks to promote Prairie immigration.
Her photographs were recently reproduced in an exhibit at the Canadian Museum of
Civilization in Ottawa, Canada.

Depictions of exposed and accessible tar deposits must have had a powerful
effect on the historical imaginations of all levels and kinds of publics in the U.S.,
Britain, and eastern Canada – especially government administrators. This effect
stemmed from the historically specific perception, as Schwartz argues, that camera
images were a mirror of reality: "the society which produced and consumed these
images placed unwavering faith in the truthfulness of the photographic image and
its ability to act as a surrogate for first-hand seeing," the photograph becoming a
"surrogate for first-hand observation – a convincing visual experience akin to 'being
there'" (Schwartz 2003a:113).

Fig. 3.4 This 1908 photograph of Von Hammerstein's drill works, entitled "Oil Derricks on the Athabasca," appeared in Agnes Deans Cameron's *The New North: Being Some Account of a Woman's Journey Through Canada to the Arctic*, 1909 (available online at http://www.gutenberg. org/files/12874/12874-h/12874-h.htm#img0104)

By presenting cornucopian images of the Canadian North, members of the Geological Survey, Von Hammerstein, and Agnes Deans Cameron partook in a process that Schwartz calls turning "sights into sites," that is, a process whereby images become cultural or symbolic shorthands that instilled preferred meanings onto a geographic space. Many books and articles about Alberta's north start with images that presage new economic staples and industrial futures (Fig. 3.5). As with other Canadian staples, the tar sands also promised enormous wealth and personal power for those who controlled it (Huber 2008). The next stage in this process, enabled by these early messengers, was to bring such visions to fruition through science.

The Scientists Arrive

Sydney Ells, considered by some "the father of Alberta bituminous sands research" (Ells:101), was clearly a man of the new frontier (see Figs. 3.6–3.7). Described as difficult and controversial, he spent over 30 years as surveyor, cartographer, and engineer researching tar sands for the Canadian Government Department of Mines.

His field survey of the bituminous sands along the Athabasca River, in summer 1913, examined reaches and tributaries south and north of Fort McMurray, first presented in a Preliminary Report published in 1914 that comprised over 85 dense pages, including 40 photographs.

In 1962, the Canadian Department of Mines published his memoir, "Reflections of the Development of the Athabasca Oil Sands" – a less technical narrative of his career in the region, covering the years between 1913 and 1945. From opening poem to closing words, Ells sketches a manly frontier where skilled "white" engineers supervised urban tenderfoots, with the assistance of shiftless Native labor. Ells' memoir includes fascinating images and descriptions of quarrying techniques, use of explosives, power shovels, and shalers in early mining activities, alongside discussions of various paving tests and results of trials at separation of the oil from the sands. Described by his colleagues as "one of these hardy pioneers," and a "rugged individual," Ells admits he was seduced by the "romance of mining" (Ells 1962:4). Despite the dense topic, his reports belie both his scientific background and purpose, coming alive with recounts, in muscular tone, of camping in the open

Fig. 3.6 Photo A12023.
Sidney Ells, 1928 appears
courtesy of the Provincial
Archives of Alberta

air at fifty-below zero, drying clothes by campfire and repairing snowshoes after a
long day of winter surveying.

Ells' tar sands images differ from the "discovery" perspectives of early pioneers
and government surveyors. He literally took the measure of the place, concentrating
his geologist's eye on verticality and not on surfaces. For months in 1913, he sur-
veyed the region, measuring the depth of bands of bitumen deposits, and the depths
and densities of its overburden. He identified high-grade deposits (many in commercial play today) and his map work is especially attentive to the commercial challenges of moving the bitumen to market, describing the locations of prime deposits,
transportation challenges, and cost predictions per ton based on distances from mine
deposit to future railheads. Ells' original 1913 maps, and a later series of topographi-
cal maps he completed in the early 1920s (surveying over 1,240 square miles of
deposits), became the only comprehensive maps of the region. One oil company
official argued that industry specialists relied upon them well into the 1950s.

Fig. 3.7 Glenbow Archives. PA 574-1039. Ells on *right*. 1945, George Sherwood Hume Fonds 1920–1957

While engaging, and at times poetic, the reports and memoir are largely devoid of attention to nature and ecosystems. Natural systems (climate, local foods, disease, weather, terrain, and more) are subtly classified as either support or inhibition to extractive strategies. Nature was not denied, but circumscribed. At best the muskeg, forest and climate are presented as human trials, obstacles to be conquered: "A fly-infested country, of many streams, in timbered or burned out areas, and almost limitless muskeg" (Ells 1962:10) to be overcome by hardy men charged with developing a modern industrial nation. Under the dual gaze of commerce and government, territory became constructed into a commodity frontier, understood in terms of deposits of natural resources, relationships to markets, and obstacles to extraction for human use. Rivers were seen for their navigable properties, not as natural ecosystems – as means to move people and technologies inland or move products out. Forests and muskeg became reduced to obstacles – not sources of biodiversity and habitat. The original inhabitants simply became a part of that landscape. Native people, who lived with a sense of the natural landscape as habitat, food, and livelihood and spirit, became a potential labor force or simply disappeared into the wilderness terrain (Sandlos 2003). Aboriginal people are glimpsed occasionally in the dotted lines on Ells' maps around reserves, or in diary descriptions of their labor as trackers and freighters hauling the tar sands south. But like most reports of the age, they are empty of Aboriginal land uses – an omission that Mitchell calls a "social hieroglyph" of the historical social relations they conceal (Mitchell 1994:10, 15).

The emphasis by Ells and others on nature as an "aid or obstacle" to resource extraction and transport silences other qualities of nature.

Ells, the engineer, blended physical masculinity with a confidence in the technological domination of nature. Such attitudes were common among engineers at the time, part of their civilizing mission. Early ecologists shared in this ideology of technological progress in the 1920s, convinced of the role of scientific knowledge in helping human society overcome the constraints of nature (Lecain 2009:57–59). In his 1913 Preliminary Report, Ells had been cautiously optimistic about the amount of oil in the region. In his memoirs, he reflects with pride on his initial estimates of the Mildred Lake-Ruth Lake deposit, noting that the later corrections to some two billion tons confirmed his early forecasts of "10 billion barrels of oil" (Ells 1962:73). Despite Ells' predictions, the scale of tar sands operations remained limited for many decades. The arrival of the railway to Fort McMurray in 1926 made some things easier. But the resource itself was not quite so free and easy as first depicted. Commodification of the resource required new methods of separation of oil from the sands that would take decades to evolve, dampening this early enthusiasm.

Karl Clark: Geographic Sites of Science

The transportation costs to move heavy and unwieldy bituminous sands to market heightened the need for an industrial process to convert bitumen into a liquid form that could flow southwards for commercial use and profit. But bitumen's special material conditions – it was dense, heavy, and mixed with sand, water, other chemicals, and clays – meant that large investments in science would need to be injected for an extensive period of time before any returns on those investments would begin to flow. No process to separate the oil from the tar sands appeared to work well at an industrial scale, although many images (below Fig. 3.8) record homespun efforts at separation. One of those whose perseverance would prevail is Dr. Karl Clark, a University of Alberta scientist and employee of the provincially supported Research Council of Alberta (founded in 1921 with the initial purpose of pursuing industrialization of the Athabasca tar). Clark is credited with developing a hot water separation method in 1926. Clark's separation research was protracted and messy. Over many years, he moved his research back and forth from university laboratory to wilderness workshop (Figs. 3.9–3.11).

David Livingstone (2003), in *Putting Science in Its Place*, argues that the location of where science is carried out adds authenticity to its claims. Images of Clark's work in the field brought increased authority to his laboratory science at the university and to the government's Alberta Research Council, and vice versa. The first pilot plants were built in Edmonton and the Dunvegan rail yards in 1924 and rebuilt in 1929 (Ferguson 1985:191 and 53–54). At the behest of the Premier and a new joint Federal and provincial Bituminous Sands Advisory Committee, Clark then established a plant in Fort McMurray on the banks of the Clearwater River in late 1929 (Fig. 3.12).

Fig. 3.8 Glenbow Archives. NA-1142-6 interior of shack with sample of tar sands and extractions.
Daniel Diver. Fort McMurray 3 March 1920

Fig. 3.9 Glenbow Archives ND-3-4596b

Fig. 3.10 Glenbow Archives ND-3-4596a Dr. Karl Clark. University of Alberta Tar Sands Department, 1929

In this early period, while the end goal was commercial investment and profit, the scientific challenge itself appeared to be a strong source of enthusiasm. Barry Ferguson argues that "the fact that the Premier himself was chairman [of the Research Council of Alberta] and that the board included both cabinet ministers and the University President" indicates its importance (1985:52). Whether intentional or not, scientific and technological problem-solving brought a dose of heroism to the process that often obscured the ultimate ends – which were always commercial exploitation. Images of bush laboratories, replete with boreholes and bubbling vats of tar, confirmed that lab experiments and processes could be adapted to the terrain and climatic conditions. This visually conveyed presence on the land was crucial to legitimating Clark's work and to attracting future commercial investment to the tar sands. Separation was shown to work in nature, that is, in "real" conditions albeit at a moderate scale. Field experiments brought with them the powerful authority of university science and suggested that underlying the separation process were scientific principles of chemistry. This lent credence to the enterprise, turning the landscape into an object of scientific intervention and privileging experts in the process

Fig. 3.11 Glenbow Archives ND-3-4596c Dr. Karl Clark University of Alberta, 1929

of unlocking and releasing a resource. Moreover, Clark's strong presence increased public awareness and acceptance, which legitimated decisions by the Alberta government to use tax payers' money for research and infrastructure, while at the same time ensuring eventual private investment in commercial-scale production (Ferguson 1985:31–58) (Figs. 3.13 and 3.14).

Industrialization

Dominion and Federal interests in the tar sands would eventually come into conflict with Alberta's over the years, as the province grew into a strong and autonomous state. The Alberta government has been steadfast in its support for development of the industry, a perseverance that has created a sense of entitlement and proprietary control that would challenge Federal jurisdiction. Jurisdiction over natural resources was officially transferred to the provinces in the 1930s, although the Federal

Fig. 3.12 Photo A 11232 oil sands extraction plants – Edmonton Dunvegan Yards, Scientific and Industrial Research Council of Alberta, 1924–1925. Courtesy of the Provincial Archives of Alberta

government maintained ownership of a leasehold and small tar sands mine at Ell's urging. The tension between the two levels of government over control of the resource continued well into the postwar period.

Between the 1930s and the 1940s, two important, privately funded commercial projects – International Bitumen followed by Abasand Oil – emerged and failed. But images of these pioneer commercial plants are sprinkled through Canadian history books, invoking pride for the industriousness of pioneers on the oil sands frontier. Robert Fitzsimmons' efforts in the 1930s are described as amateur and promotional, but they also caused a contagious "fever of belief" among his followers. Ferguson describes the hypnotic effect of the tar sands on engineers and scientists who worked with Fitzsimmons to try to make separation and production commercially viable, but by 1939, just 9 years after construction, the International Bitumen plant was considered "worthless" and consigned to the dustbins of industrial history (Ferguson 1985:85) (Figs. 3.15 and 3.16).

The Abasand Oil plant began operations in 1936 and was considered a vast improvement over the first such effort, constructed under the guidance of Max Ball,

Fig. 3.13 Photo A 3344 PAA oil sands – ½ mile north of Fort McMurray – being examined by
Alberta Premier Herbert Greenfield and others, 1921 appears courtesy of the Provincial Archives
of Alberta. *Right* to *left*: S.E. Mercier, Northern Construction Company; Hon. V.W. Smith, Minister
of Railways; Premier Greenfield; Col. Jim Cornwall, Northern Transportation Company (Holding
easily available tar and contemplating its future uses appears over and over in the visual record.
Here the powerful hands of politics and commerce.)

Fig. 3.14 PAA 11223 oil extraction plant – Clearwater River plant ca. 1930 appears courtesy of
the Provincial Archives of Alberta. The plant processed 800 tons of oil sands in the summer of
1930, and yielded over 75 tons of bitumen (Ferguson 1985:54)

Fig. 3.15 PAA A3384 International Bitumen Company 1930 early process of liquefying Bitumen appears courtesy of the Provincial Archives of Alberta

Glenbow Archives NA-3394-57

Fig. 3.16 Glenbow Archives NA-3394-57. Bitumount Tar Sands Plant 1936 Fitzsimmons

an American engineer. The Abasand plant took years to develop and test, and images of it give the impression of a professional and technically sophisticated industrial enterprise. The plant began to produce petroleum products of various kinds in 1941, but was destroyed by fire in November of that year, never reaching its "design capacity of 3,000 barrels of bitumen per day," reaching on average 400 barrels of oil per day

Fig. 3.17 Glenbow Archives. PA-574-1077 Abasand plant. G.S. Hume collection, 1944/1945 album

in the summer of 1941, although it did prove extraction could be effective (Ferguson 1985:204). Two key factors were common to these commercial ventures: "they followed paths beaten by government researchers," like Clark, and they required "advanced technical expertise and large sums of money if commercial development was to be successful" (Ferguson 1985:94).

During early industrial field trials of 1920s and 1930s, and again during World War II, growing fears of the scarcity of oil supply boosted enthusiasm and support for research and development. The use of energy security discourse is a common ideological frame employed by proponents at the time, one that would reappear in the 1970s during first oil crisis, and once again in the contemporary post-9–11 discourses, with narratives in the latter period infused with energy and political security alike. Faced with fuel shortages during World War II, at a time when Canada produced only five percent of the oil it used, the federal government decided to resurrect the failing Abasand project and started reconstruction in 1943 (Comfort, 1980) (Figs. 3.17–3.19).

A number of federal investigators carried out a series of studies to identify problems with Ball's engineering, plant design, and operations, and the technical changes needed to guarantee commercial potential of the tar sands, including the need to upscale the plant in size (Ferguson 1985:93). The Federal Department of Mines, with Sydney Ells at the helm, took the lead financial role, although some private Canadian money remained in play. They moved the operation to Mildred Lake on a leasehold controlled by Ottawa (originally scouted by Ells) containing richer deposits

Fig. 3.18 Glenbow Archives. PA-574-812 Abasand Plant. G.S. Hume collection, 1942 album

of tar and expanded the project to 4,000 barrels of bitumen a day. Alberta was wary of the project, fearing a federal takeover of the resource. Various authors report the lack of collaboration between federal and provincial specialists and scientists, including Ells and Clark. In the end, the project was a spectacular failure, and again burned down in 1945. According to Chastko, "the simmering conflict between the two groups allowed the oilsands issue to become subsumed within the federal-provincial battleground over natural resource development" (Chastko 2004:54).

Bitumount: Success at Last

In an effort to maintain the interest of private investors in the tar sands, the Alberta government funded the building of Bitumount, a new experimental separation plant in 1946 and 1947, near the site of the old Fitzsimmons plant. A large-scale pilot

Fig. 3.19 Glenbow Archives. PA-574-1074 Abasand plant. G.S. Hume collection, 1944/1945 album

plant, it was redesigned by American engineers under the supervision of Clark's colleague Sydney Blair; its purpose once again to prove commercial viability. The plant operated for two seasons, beginning in the summer of 1948, and intermittently thereafter for another decade, at a total cost to Alberta taxpayers of about one million dollars over the lifetime of the plant (Ferguson 1985) (for comparison, the entire 1953–1954 Alberta budget was $53 million). Employing Clark's hot water method, it only produced 500 barrels of bitumen a day, but "proved viability of the separation process and bitumen production" (Ferguson 1985:209–211) (Fig. 3.20).

The Government allowed private firms such as Can-Amera Oil Sands Development, Royalite Oil Company, and Great Canadian Oil Sands (GCOS) to run their own tests in the plant. And in 1950, Sydney Blair conducted a cost analysis for the Alberta government that was recognized as the single most influential analysis of tar sands profitability. He concluded that the scale of any future commercial plant would need to be larger than 20,000 barrels/day to be profitable. He also noted that considerable waste ponds would need to be created at this scale. Particularly in light of this analysis, and a constituency in conventional oil that was growing in strength, the Alberta government's enthusiasm became tempered with new caution: "the oil sands were an expensive alternative fuel operating on the margins of the oil industry...the province continued to worry that development of the oil sands could harm the interests of conventional producers" (Chastko 2004:103–104). Chastko details

Fig. 3.20 Glenbow Archives. PA-1599-451-2 aerial view of provincial government's pilot plant for extracting oil from northern Alberta tar sands. Bitumont, Alberta C, 1949–1950. In 1974, the Bitumont site, located 89 km north of Fort McMurray, was designated an Alberta historical site. See the statement of historic significance at http://www.historicplaces.ca/en/rep-reg/place-lieu.aspx?id=4998

the Alberta government's unwillingness to kill the "golden goose" of Alberta's conventional oil industry with competition from the tar sands.

Chastko describes the "reluctant expansion" of the tar sands in the 1960s as the Alberta government negotiated with competing American investors. Sun Oil and its President J. Howard Pew would eventually take the lead to develop the first commercial operation – the GCOS project, now Suncor. Sun Oil purchased the Abasand Oil lease at Mildred-Ruth Lakes and made a deal with GCOS in 1958 to use technology developed by that company. In order to guarantee a market for the product, Sun prepurchased 75% of the output at a contracted price, at a time when prices were expected to rise, as projected US and Canadian conventional oil production appeared to exceed projected demand. Just when the tar would finally express its full commercial potential, Alberta's Premier Ernest Manning attempted to protect the conventional oil industry by capping tar sands output at no more than five percent of Alberta's annual conventional oil production, a goal his Government adopted in 1962. Manning's ceiling thus restricted the scale of the entire tar sands industry to less than 30,000 barrels. It also included plans for stiff lease and royalty payments

Fig. 3.21 Glenbow Archives S-229-21 heavy machinery used at Athabasca Tar Sands, 1967

of "8 percent on crude production up to 900,000 barrels and 20 percent" thereafter, as well as 16.66% on by-products (Chastko 2004:115).

Sun Oil hired Canadian Bechtel engineering to reexamine the feasibility of a smaller plant. To the contrary, Bechtel's engineers concluded that a plant of about 45,000 barrels a day was required, more than double Sydney Blair's earlier conclusions. Private capital, convinced by decades of state endorsement of tar sands development, now ironically faced the need to present their business case to a reluctant Alberta government. The proposal, ultimately endorsed, saw Sun Oil increase their ownership share of GCOS to 83% in exchange for their providing financial support for the larger plant. Manning and the Alberta government agreed to the increased scale (about 7.5% of the overall oil output in the province), despite protests from Sun's competitors and not withholding Alberta's own misgivings about finding new markets for both conventional and synthetic oil producers (Figs. 3.21 and 3.22).

Once the ink was dry, the GCOS project finally opened to much fanfare in 1967. Visuals of large, commercial-scale plants such as Abasand and Bitumount in the 1940s and 1950s, and the even larger GCOS in the 1960s, further encouraged private and public enthusiasm. While the size of these plants is miniscule compared to today's scales, the visual impact of professionally engineered production plants on the edge of the boreal wilderness invoked paradigms of progress and human domination over nature more than words ever could, especially when contrasted with early images of boiling pots of oil and slipshod systems of mining and separation. The 1949 photo of Karl Clark (below) at the modern Bitumount pilot oilsands plant

Fig. 3.22 Glenbow Archives S-229-26 aerial view of Athabasca Tar Sands Plant, 1967

depicts a gentleman scientist, in his comfortable old sweater, pipe in hand, standing in front of the fruits of his research. Clark looks more like he has just risen from his easy chair beside a fireplace, providing a certain public reassurance that the tar sands have arrived, nature at last tamed by science and industriousness. He would attend the sod turning of the GCOS in 1965, but died some months before it went into operations. Mary Clark Sheppard would write in her biography of her father (1989) that Karl Clark was a muscular Christian engineer with a deep respect for nature who, at the end of his life, was shocked to see the gargantuan scale of mining and processing required at the GCOS plant. (Of course, even this scale was fledgling compared to today's production of nearly 2 billion barrels/day) (Fig. 3.23).

Continuities: Visual Threads, Ideological Mainstays

This chapter introduced a series of historical tar sand images to illustrate how representations of resource exploitation supplemented the state's endorsement of such a speculative undertaking: in essence, the normalization of mass destruction. Many of these older images still circulate today, representing more than historical sketches of the resource and its geographic space; they offer a continuity of narrative about nature's stubborn resistance, matched by an equally stubborn and ultimately prevailing human ingenuity. Technology overcomes nature through innovation and

Fig. 3.23 Photo PA410.3 "Dr. Karl Clark who has been in charge of the bituminous sands project since 1920" appears courtesy of the Provincial Archives of Alberta

determination by scientists, entrepreneurs, and government officials, all to the fanfare of citizens. This visual narrative, presented as a factual photographic index of the step-by-step development of the industry, has crystallized into a set of resilient and iconic metaphors used by corporate and state interests, perhaps displayed most suc-cinctly by the curators of the industry-supported Oil Sands Discovery Center in Fort McMurray. Such visual narratives enhance state legitimacy, infuse popular imagi-naries, and are today celebrated as Alberta heritage and culture: "the same photograph circulat[ing] in time and space, between historical document and formal experiment, from mass media to curated exhibition" (Gabara 2006:140, 167).

Historic myths fuel contemporary perspectives. Tropes of the heroic struggle by masculine scientists and private entrepreneurs to release the bounty of the north, and the nurturing role of the state, continue to embellish political discourse today.

The seeming overnight, free market-induced appearance of the tar sands is juxtaposed with a healthy dose of irony upon this history, in which the same narratives that are today invoked alongside neoliberal fears of the presumed catastrophic consequences of state involvement offered equal support for an enterprise that is far more aptly defined as state than private. According to Ferguson, "Alberta's push for development took priority over its commitment to private leadership" (1985:147). The next chapter explores the story of Syncrude and the transition to full-scale production in the contemporary industrial period.

The mass production of diffuse deposits of ores (like the tar sands) requires mass destruction (Lecain 2009), including huge, destructive open pit mines, the removal of hundreds of thousands of tons of overburden, enormous amounts of energy for separating and processing the materials and more, just to derive small amounts of actual ore or, in our case, bitumen. LeCain argues that mining sites create a "simplified environment" where "nature still very obviously exists, but it exists as part of a hybrid ecological and technological mining system," ..."a constructed human environment, created by experts (engineers, managers, skilled miners) through theory knowledge and limited mastery of subterrestrial space and ecology" (2009:21). For LeCain, the key property of open-pit mining is not only "its sheer destructiveness," but also the speed at which it is occurring (2009:8). The scale is "so big as to defy easy description... literally beyond the normal human sense of scale and proportion...People simply grasp for analogies to the biggest thing they have ever heard of " (Lecain:11).

The normalization of such mass destruction might not have unfolded in quite such uncontested terms were it not for the unique Western celebration of all things grand and grandiose. This process of mass destruction required to cheaply extract "huge amounts of essential industrial minerals from the earth's crusts" has been largely hidden "from urban worlds of mass production and mass consumption" unrecognized for the most part by a society which revels in the jobs and the lifestyle benefits of energy (LeCain 2009:7). According to John Berger, "Nothing is accidental in an image," and visual depictions about tar sands development convey scale, perhaps more than any other message (Fig. 3.24). If Karl Clark in his later years expressed alarm at the scale of destruction that ensued, his concern was easily drowned out by the "Wow" factor expressed by visuals of the massive machinery in use. Government and commercial promoters of the tar sands industry in the 1960s would celebrate this immensity, and images like the bucket-wheel, shown above dwarfing human figures, became a selling feature to the public, symbolizing the enormity of the challenge they had overcome. This visual record was and is disseminated alongside a discourse of immensity of the resource itself. Scale would bite back eventually, becoming a public relations liability for companies in contemporary times (LeCain 2009).

But in its heyday, the bucket-wheel would become an icon for the new industry, reproduced by GCOS on commemorative medals in 1975 (on the tenth anniversary of public debentures purchased by over 100,000 Albertans to sustain the company), and in a stamp by Canada Post honoring the industry in both official languages in 1978 (Figs. 3.25 and 3.26). Raento and Brunn (2005) consider the visual qualities of stamps much like social texts that offer "readings" of a territory or project. Stamps can act like political messengers, and their mundane omnipresence has considerable

Fig. 3.24 Glenbow Archives. S-229-21a. Heavy equipment used at Athabasca Tar Sands, 1967

Fig. 3.25 Pioneering energy together. Great Canadian Oil Sands Medallion, 1975

Fig. 3.26 Library and Archives Canada; Copyright: Canada Post Corporation. MIKAN 2218481, 1978, with permission

nation-building power. Today, the bucket-wheel has been retired to the outdoor museum at the Oil Sands Discovery Center in Fort McMurray, having been surpassed by new technologies that make use of even larger equipment. But it served its purpose. With the enthusiasm surrounding the success of the GCOS, its *scale of production* was celebrated as much as the *scale of destruction* remained hidden – or more accurately displaced – by the wonders of mega technology.

Today, with the growing democratization of visual discourse, and in an era of growing ecological alarm, the discourse of immensity has become unsettled by critics like Greenpeace who describe the scale of ecological disturbances of the Athabasca tar sands, which has a footprint "larger than the state of Florida."

"Big" is still big in politics though, with images like bucket-wheels, drag lines, immense earth moving equipment, and today the ubiquitous giant truck still being used to inspire awe and political support. Giant equipment is often juxtaposed with tiny humans as a visual trope in which the harm caused to nature by gargantuan-scale mining is minimized, tearing up the boreal forest becomes likened to big boys (and a few girls) with big toys playing in a sandbox. Industry does not hide its operations, but displays them in plain sight, celebrating scale in front and center. Today, the large truck has become the major trope (Fig. 3.27).

In Washington DC in 2006, while then Premier Ralph Klein and his Cabinet Ministers met with the Vice President, Congressional representatives, senior Washington bureaucrats, and the Chairman of the Senate Committee on Energy and Resources, across the mall at the *Alberta at the Smithsonian Exhibit* the province displayed an oil sands haul truck for show and tell. Both the truck and the tires are

Fig. 3.27 September 2005 , 08:20. Caterpillar 797 truck empty weight: 557,820 kg (1,230,000 lb). Drive: 3524B EUI twin turbocharged and after cooled diesel engine. Max speed: 64 km/h or 40 mph. Horse power: 3,500. Suspension: self-contained oil pneumatic suspension cylinder on each wheel. Height empty: 7.1 m (23 ft 8 in.), length: 14.3 m (47 ft 7 in.), body width: 9.0 m (30 ft). Dumping height: 14.8 m (49 ft 3 in.). Tire size: 3.8 m (12.9 ft) in diameter. Cost: $5–6 million Creative Commons. See also Peter Essick, http://s.ngm.com/2009/03/canadian-oil-sands/img/candian-oil-sands-truck-160.jpg

built in the USA, evincing Canada's economic dependency on the U.S.[1] Meanwhile, back at the Fort McMurray Oil Sands Discovery Center, visitors find the theme reinforced in the Gift Shop's offerings of dinky toys, truck and shovel books, and souvenir clothing. Mass destruction becomes normalized in this reconstruction of meaning. Images often resemble toys, blinding readers to the destruction required by mining and operations, and minimizing the challenges of reclaiming and restoring the massive disturbances to natural systems (Fig. 3.28).

The dominant, progressive interpretations of gargantuan earth moving equipment are today contested, representing progress to some, the destruction of nature to others. In the following chapters, we trace the shift in visual messages as the exhaustion of nature and the mass destruction of ecosystems required to provide energy

[1] http://www.innovation.gov.ab.ca/general/tra_exp/Final_overview_Alberta_mission_to_Washington.pdf. Accessed January 7, 2011.

Fig. 3.28 My new socks from the Oil Sands Discovery Center. Photograph by M. Gismondi, December 2009

and fuel for our lifestyles has entered the public consciousness. But, as with many social problems, we allowed this particular problem to get much, much worse before we began to contemplate making it better, and "Big" took on a whole new meaning in the decades since the retirement of the bucket-wheel, an era that began with the Arab oil embargo of the 1970s, legitimating the exponential expansion of tar sands development by a new corporate player, Syncrude.

References

Chastko, P. (2004). *Developing Alberta's Oil Sands from Karl Clark to Kyoto.* Calgary: University of Calgary.

Comfort, D. (1980). *The Abasand Fiasco: The Rise and Fall of a Brave Pioneer Oil Sands Extraction Plant.* Edmonton: Friesen Printers.

Courrier de l'ouest. (1910). Une Nouvelle Compagnie de Petrole. 13 January:1. Oct. 1905-Dec. 1916 (French Language Newspaper).

Deans Cameron, A. (1909). *The New North. Being Some Account of a Woman's Journey through Canada to the Arctic.* New York: Appleton.

Ells, S.C. (1962). *Recollections of the Development of the Athabasca Oil Sands.* Ottawa: Department of Mines and Technical Surveys, Mines Branch.

Fairclough, N. (2009). Language and globalization. *Semiotica 2009* (173), 317–342.

Ferguson, B. (1985). *Athabasca Oil Sands: Northern Resource Exploration 1875–1951.* Regina, SK: Canadian Plains Research Center.

Gabara, E. (2006). Recycled photographs: moving still images of Mexico City, 1950–2000. In Schwartz, M. & Tierney-Tello, M.B. (Eds.). *Photography and Writing in Latin America: Double Exposures*. Albuquerque: University of New Mexico.

Huber, M. (2008). Energizing historical materialism: fossil fuels, space and the capitalist mode of production. *Geoforum* (40):105–115.

Lecain, T.J. (2009). *Mass Destruction: The Men and Giant Mines that Wired America and Scarred the Planet*. New Brunswick, NJ: Rutgers University.

Livingstone, D. (2003). *Putting science in its place: Geographies of scientific knowledge*. Chicago: University of Chicago.

Mitchell, W.J.T. (Ed.). (1994). *Landscape and Power*. Chicago: University of Chicago.

O'Leary, D. (2004). Environmentalism, hermeneutics, and Canadian imperialism in Agnes Dean Cameron's *The New North*. In Hessing, M., Raglon, R., & Sandilands, C. (Eds.). *This Elusive Land: Women and the Canadian Environment*. Vancouver: UBC.

Sandlos, J. (2003). Landscaping desire: poetics, politics in the early biological surveys of the Canadian North. *Space & Culture 6*(4), 394–414.

Raento, P. & Brunn, S.D. (2005). Visualizing Finland postage stamps as political messengers. *Geografiska Annaler Series B: Human Geography 87*(2), 145–164.

Roy, W. (2005). Primacy, technology and nationalism in Agnes Deans Camerons' *The New North*. *Mosaic (Winnipeg) 38*(2), 53–79.

Schwartz, J. & Ryan, J. (2003). Introduction: Photography and the Geographical Imagination. In J. Schwartz and J. Ryan (Eds.), *Picturing Place: Photography and the Geographical Imagination*. London: Tauris & Co.

Schwartz, J. (2003a). More than 'competent descriptions of an intractably empty landscape': a strategy for critical engagement with historical photographs. *Historical Geography 31*, 105–30.

Schwartz, J. (2003b). Photographs from the edge of empire. In Blunt, A., Gruffudd, P., May, J., Ogborn, M., Pinder, D. (Eds.). *Cultural Geography in Practice*. London: Arnold.

Zaslow, M. (1971). *The Opening of the Canadian North 1870–1914*. Toronto: McClelland and Stewart.

Zeller, S. (1987). *Inventing Canada: Early Victorian Science and the Idea of a Transcontinental Nation*. Toronto: University of Toronto.

Chapter 4
Capital, Labor, and the State

> *I don't need to tell you that Fort McMurray is the engine that*
> *drives much of Alberta's and Canada's economies. The world*
> *will be looking more and more to Alberta and Fort McMurray*
> *for energy in the years to come (Albert Premier Ed Stelmach,*
> *Fort McMurray Chamber of Commerce, Fort McMurray, 24*
> *Jan 2007).*

Incumbent Alberta Premier Ed Stelmach (2006–) says that he "doesn't need to tell" the Fort McMurray business community how their support for the Athabasca oil sands is key to the economies of Alberta and Canada. He then announces to citizens of the city and the province that it is their responsibility to supply energy to the world. Much government and industry rhetoric about tar sands development frames facts in the form of proclamations: barrels of oil produced, capital invested, jobs created, royalties paid, and global obligations. Government accounts emphasize how politicians are managing the oils sands as a public resource, in conjunction with citizens; the results and the burdens shared equitably for the public good. Huge corporate profits are seldom mentioned, nor are the actual government and corporate decision-makers revealed. Rather, decisions and actions driving oil sands investment and expansions are presented as abstract forces, such as global consumer demand for oil, the free market, and economic globalization. The role of government, if there is one, is to nurture a fledgling but vital industry, by providing infrastructure and economic subsidies to meet a unique investment "moment." In exchange, global corporations will develop the crown resources in the interest of the public. Albertans, and especially outsiders, dare not interfere with the market (or the pace of development) because oil sands growth ensures Canada's place in the global economy. Indeed, the world is depending on the people of McMurray.

Reality does not always support such privileged accounts. Global corporations and shareholders are seeking profits, not energy, from the tar sands. And, despite a global Who's Who of corporate investors, the only market for tar sands oil currently is the United States, not the world. In fact, one-fourth of the energy used in the United States comes from Canada (Pasqualetti 2009), although most of this is from conventional sources. Moreover, while capital may well move at the speed of light, workers and the

D.J. Davidson and M. Gismondi, *Challenging Legitimacy at the Precipice*
of Energy Calamity, DOI 10.1007/978-1-4614-0287-9_4,
© Springer Science+Business Media, LLC 2011

tar sands themselves cannot be so readily digitized. The very materiality of extraction, upgrading, and refining requires investment capital, energy and water, equipment, labor and the associated food, shelter and services to reproduce a mostly mobile workforce in a harsh northern frontier, and competently governed urban centers with industrial and financial services. To process and move oil out to the world for consumption requires yet more capital investment, energy, fluids, pumps and pipelines, as well as more labor to construct and maintain distribution systems. A different story about the tar sands is told by its "place" within these systems of global flows, themselves legitimated by complex ideological frameworks and key messages from government leaders. In contrast, in this chapter we explore evidence of privileged access to the benefits and profits by Canadian and transnational corporations, and evidence of risks borne in disproportion to gains by ordinary Albertan taxpayers, labor, the communities like in Fort McMurray, and Aboriginal peoples groups downstream, and, if greenhouse gases are considered, future generations around the globe.

It is easy to see why the privileged accounts offered by government and industries are persuasive. The world price for oil tripled between 1991 and 2006 from US $26 to 76 per barrel, and touched $147 per barrel in July 2008. The flow of global capital into the region is staggering (Clarke 2008), and a line-up of leading global transnationals have purchased oil sands leases: Shell, ConocoPhillips, Synenco/Sinopec, British Petroleum, Exxon, Imperial Oil, Total, StatOil, and many more (see oil sands lease map, available at http://www.opticanada.com/projects/oil_sands_overview; http://www.energy.alberta.ca/LandAccess/pdfs/OilSands_Projects.pdf). In the government's *Talk About Oil Sands Fact Sheet*, chatty writers communicate the good news: an outlay of more than $91 billion between 1999 and 2009 (Alberta Government 2010a, b). With prices at over $100 per barrel in 2008, the *Fact Sheet* reveals Alberta Government calculations that "oil sands investment reached a record-high $19.2 billion, a 14% increase over the 2007 level," and that Alberta collected "$3 billion in royalties from oil sands projects" that year alone. When prices declined to $40 per barrel in late 2008, the *Fact Sheet* writers calmed the waters with confirmation of reinvestment "some $10 billion or more in 2009," as well as "already planned or in process" another "$170 billion of new investment" for 2010 forwards. The authors close with claims that "42,000 new jobs will be made in the next 20 years." Oil sands investment equals jobs is the message; one rosy forecast predicts 450,000 jobs across Canada annually for the next quarter century (Canadian Energy Research Institute 2009).

Estimates vary, but according to current estimates, oil companies are producing 1.5 million barrels of synthetic oil a day in the tar sands. Alberta's Energy Resource Conservation Board predicts that production will reach 3.2 million barrels a day by 2017 (AERCB 2008). As well, the percentage of bitumen that will be upgraded in Alberta has increased to 59% (Sinclair 2011:71), although this appears in decline (and far below promised government targets of 75%) as upgrading is reportedly being exported to the United States (Calgary Herald 2008), in part due to the global recession. A slowing in the pace of project development has resulted in some delays, but many projects are proceeding full steam ahead, and the long lists of corporations that have since acquired new leases and committed to mines and in situ project expansions suggest that the boom will soon be full-on again (Strategy West Inc. 2009). The provincial government presents future expansions as unproblematic, in ways that suggest everyone wins from such huge investments. But even the Mayor of the Regional Municipality of Wood Buffalo (encompassing Fort McMurray) Melissa Blake (2010) described the recent slowdown as a "welcome pause" and "a period of relief from the last two years." During the years preceding the recession, the impacts of the boom led the regional municipal government to take the unprecedented step of strategically intervening in a series of new project reviews by the Alberta Energy and Utility Board, insisting on additional provincial to help "to cope with gaping infrastructure needs" (Edmonton Journal 2006). While the Mayor remains an unstinting supporter of the energy industry, she does not mince words when identifying the social and infrastructural pressures in the Regional Municipality of Wood Buffalo caused by rapid industrial expansion over the last decade (Blake 2010).

Other critics, including global and local public citizens, environmentalists, opposition politicians, the Alberta Federation of Labour (AFL), as well as unlikely voices such as former Conservative Party leader and Alberta Premier Peter Lougheed – the mastermind behind the 1980s tar sands boom – argue that not everyone outside and inside the oil sector is enjoying the fruits of this rapid pace of development. They report wage and price inflation, housing shortages, infra-structure deficits, increased health risks, labor market distortions, spiraling real estate costs, and community and family disruptions. Some question why projects remain heavily subsidized by Albertan and Canadian taxpayers, including govern-ment funding directed to civic infrastructure of direct service to the industry (high-ways, airports, and bridges servicing the tar sands operations) instead of support for health care and education for Albertans. Others wonder why royalties are so low, after all, other oil regions of the world demand far higher royalties from the same companies, and tax breaks are unnecessary for a mature industry. Still others have questioned why Alberta's Heritage Fund ("Alberta Heritage Savings Trust Fund," the public savings plan that relies on deposits from royalties and lease charges paid by oil companies) is much smaller than those in Alaska or Norway? In *Misplaced Generosity,* Boychuk (2010:20) confirms public suspicions that the concentration of benefits are flowing to transnationals: "the companies operating in Alberta's tar sands since 1997 have exchanged $19.3 billion in royalties/land sales for $205.5 billion worth of Alberta's natural wealth," a royalty/profit ratio far below Alberta government targets, and even further below percentages paid in other countries.

Background

We did not get here overnight. Historical and political context highlight the lengthy roots of current debates, the rocky path of highs and lows in the growth of the indus-try, and the conflicts between corporations, Federal and Alberta governments, and the public. Since 1971, the Conservative Party has ruled Alberta, as some critics argue, like a one-party state. The concentration of political power for 40 years, in combination with oil revenues, has allowed Alberta's Conservative Party to walk its own path and deflect its critics, including recent Federal attempts to introduce national environmental standards and economic resource policies and taxes. As stewards of a maverick petro-state on the Canadian political scene, Alberta's Conservative provincial premiers have been adept at playing brinkmanship politics with the Federal Government in Ottawa, especially around resource and environmental issues. Even before the signing of the North American Free Trade Agreement in 1993, Alberta aggressively privatized crown resources, leasing much of the forest cover of the Athabasca region to Japanese pulp mills in the 1980s (Pratt and Urquhart 1994). In the 1990s, Alberta and Canada signed favorable royalty and tax agreements to entice more development in what was then a fledgling tar sands industry. Successive Alberta Premiers established independent relations with American oil and pipeline corporations, made links with US senators and Washington bureaucrats, and even installed their own Alberta "embassy" in Washington, D.C. (Lisac 2010). Since the turn of the millennium, Alberta's governing Tories have been effective in rallying other Canadian provinces to rethink Federal attempts to adopt pollution emission caps and the Kyoto Protocol, claiming that national standards would impact Alberta's tar sands, which, they reminded the national public, is driving the nation's economy. Critics argue that contemporary Alberta governments, like other cash-flush petro states, have become arrogant and overly self-confident (Soron 2004; Shiell and Loney 2007; Paehlke 2008).

Alberta has not always been a "have" province. The Canadian economy continues to be a staples economy and Prairie Provinces like Alberta are no exception: grain and beef, lumber, pulp and paper, and minerals; and since the early 1900s, oil and natural gas, have been Alberta's primary sources of economic development. As primary goods, most left the region for value-added secondary manufacturing and upgrading elsewhere in eastern Canada or the United States, creating more jobs and wealth outside the province than in. Historians often portray Alberta politics through the lenses of a series of historical grievances between Alberta (a new landlocked province (founded in 1905), geographically remote, resource rich and lightly populated), and a powerful federal government in Ottawa, intent on controlling Alberta's resources, and apparently under the sway of central Canadian economic elites and bankers. This depiction draws on a long series of grievances that included the federal government's free land grants to the Hudson's Bay Company and Canadian Pacific Railway (corporations that then sold the former public land back to settlers); grain farmers complaints about the railway monopoly and its exorbitant transport rates to move grain to markets; frustrations about "protective federal tariffs" that stifled free markets; and difficulties accessing credit from central Canadian banks and elites (Finkel 1989; Palmer and Palmer 1990). Many of these early grievances arose during the agrarian period in Alberta, when there was little wealth to go around.

Conventional oil was discovered in Alberta in Turner Valley in 1914, but full recognition of Alberta's substantial conventional petroleum reserves only began with the Leduc oil discovery in 1947. With the development of large oil and gas deposits over the last 60 years, Ottawa's interference in Alberta resource politics has appeared to Albertans as a blatant attempt to steal from a former have-not province that "struck it rich." Richards and Pratt (1979), however, trace current federal-provincial tensions to the building of provincial state institutions in the Prairies in the 1960s and 1970s, programs of pragmatic social policy or populism that emerged from the geographic isolation, rural self reliance, and make-do attitudes of Prairie people, describing the development of a form of economic regionalism. For 36 years, starting in the mid 1930s, the Social Credit Party had governed Alberta from a strong rural political base. As the economy began to change with the growth of the conventional oil and gas industry (mostly US-controlled firms), industry revenues and royalties paid to the state – what Owram (2005) called "Oil's Magic Wand" – allowed the Alberta government to expand or establish new programs and services, create jobs, and stimulate growth of an urban middle class. This combination of pragmatism, regional nationalism, and populist roots set the West apart from central Canada and contributed to Alberta's independent province building. It is from within this political-cultural context that Alberta's energy politics emerged.

In the 1970s, Peter Lougheed and a new Progressive Conservative party launched into provincial leadership, having gained key support from the urban middle classes that provided a willing audience for his Alberta First economic ideology, including promotion of an Alberta-owned and controlled energy industry, and government support for entrepreneurs (Pratt 1977). His plans coincided with a global increase in oil prices following the Arab–Israeli war in 1967. A boom in exploration and extraction of oil and gas across the province followed, and petroleum revenues poured in. When Alberta wanted to raise oil prices to match world prices, Canadian Prime Minister Pierre Trudeau and his Liberal government resisted in order to keep costs low for refineries and consumers in eastern Canada, as television newscasts reported lineups at gas stations, and predicted a socioeconomic crisis because of fuel shortages. Despite Lougheed's protests, Ottawa imposed a federal export tax on Alberta oil, one of several such federal acts that were perceived by Albertans as "a blatant move to siphon off cash in order to subsidize consumers and industries in Ontario and Quebec" (Owram 2005). The Lougheed government, revealingly, asserted autonomy and ownership over petro-resources to secure provincial rights to export before serving national consumers, a pattern of politics that persists today. While the progressive populist culture described here has given way to Western separatism and economic elitism, a certain cowboy ideology of frontier individualism remains, although nowadays the new Alberta cowboy is a pipeline cowboy (Filax and Specht 2009).

Lougheed and the Tar Sands

In 1973, rising oil prices (from US $3 to 11) made the capital intensive tar sands projects more attractive investments. Old siren songs of tar sand reserves "the size of Saudi Arabia" attracted an American-led investment consortium of Imperial Oil, Gulf Canada, Atlantic Richfield, and Canada-Cities Services to form Syncrude, "which sought a provincial permit to build a large plant to extract oil from the

Fig 4.1 Glenbow Archives. M-8787-169. Syncrude Canada gets stuck in the Alberta tar sands. 11 Jan 1975. Atlantic Richfield withdraws from project, $1 billion needed to help pay for increased costs in order to continue. Photographer/Illustrator: Innes, Tom, Calgary, AB

motherlode" (Marsh 2005:653). University of Alberta politics professor Larry Pratt (1976) caused a stir with his book *The Tars Sands: Syncrude and the Politics of Oil,* an unflattering reading of Lougheed's negotiations with the oil majors. Today's readers will be struck by Pratt's analyses of discourses used by politicians and oil industry elites to legitimate government support for Syncrude. Many resonate with contemporary debates: the need for security of supply in the face of Middle-East political turmoil; burgeoning United States' energy needs; American attempts at energy autonomy replete with labels like Project Independence and Fortress North America energy pool; and cartoons lampooning the political tensions between Alberta and Ottawa. Pratt discovered one concession after another amidst the ensuing negotiations, as big oil companies outflanked the Albertans, although Ottawa discouraged Atlantic Richfield with a tax plan in 1974 that reduced corporate concessions, just as construction costs hit $2 billion. Atlantic Richfield's withdrawal threw the initial deal into peril (Fig. 4.1).

By early 1975, with over 1,500 workers already on the Syncrude construction site in Fort McMurray, and suffering from a general "capital strike by the oil industry," Alberta, Ontario and Ottawa offered more incentives, took over Richfield's financial position, eventually giving Canada 15%, Alberta 10%, and Ontario 5% of the stakes, to ensure that the oil majors would stay and develop the plant. Pratt cried big time boon-doggle as governments took the bulk of the financial risks, for only 30% of the equity, another precedent that persists today.

Pratt draws in readers with his revelations of raw power at work, as the oil companies exploited divisions between Federal and Provincial governments. By 1975, Alberta (and the federal government and Ontario) further underwrote the development and the final terms became even sweeter, providing "a guaranteed rate of return on investment, royalty-free holidays, commitments to provide strike-free labour at work sites, provincial support in Syncrude's negotiations with Ottawa, expensive publicly financed 'roads to resources' infrastructure, and later on federal tax avoidance, increased provincial royalty write-downs, and more" (Pratt 1977:133–147). Pratt concludes, as did some insiders, that it was more akin to a kind of economic blackmail, where oil companies deny capital and threaten to pull out investments if they do not get government subsidies. Pratt wrote: "what has been slipped across, or under, the bargaining table in such cases is not lost 'economic rent,' but the less tangible and far more valuable asset of sovereignty – the right and power of a people to make their own choices, to design their own future" (1977:166).

In 1977, TV director Peter Pearson co-wrote and developed a special episode of *For the Record*, a Canadian Broadcasting Corporation (CBC 1977) television documentary, which depicted Lougheed and Alberta's handling of the deal. After a single showing to an audience of over one million Canadians, Lougheed sued CBC (the parties settled out of court in 1983), complaining that the docu-drama "falsely depicted him as a 'weak and irresolute' politician who had been 'out-maneuvered (and) outsmarted'" by hard-nosed oil company negotiators. Lougheed, a Harvard trained lawyer, had come to politics from the Mannix Corporation, one of the largest construction firms in Canada with family links to the Bechtel Corporation in the United States. Canadian Bechtel Ltd. constructed Suncor and would construct Syncrude, the "largest construction project of its kind in the world," in 1978 (Bechtel 1998). No one has ever explored how those linkages may have collared the Syncrude deal. Lougheed claimed CBC falsely depicted him as trying to hide from the public the nature of the "sweetheart deal" struck with the consortium (Sallot 1982). Today's public, accustomed to Wiki-leaks and public use of internet to distribute video footage embarrassing to the government, may be surprised to know that this 1977 film has never aired again and is not available from the CBC or public archives (Epp 1984) (Fig. 4.2).

The First Boom: 1973–1984

A steep rise in global oil prices in 1974 outstripped national plans to freeze Alberta oil prices, and as prices increased so did Alberta's economic rents. In 1975,

Fig. 4.2 Glenbow Archives M-8000-149. Premier Lougheed takes aim at the Canadian Broadcasting Company. Innes. Calgary Herald

Lougheed scored a resounding election win. The year before, he founded the Alberta Oil Sands Technology and Research Authority (AOSTRA) "to promote the development and use of new technology for oil sands and heavy oil production; provides seed money and funds research to make oil sands economically feasible." For more on AOSTRA see Alastair Sweeney's Black Bonanza Timelines (http://www.alastairsweeny.com/blackbonanza/index.php/Black_Bonanza_Timeline_-_Oilsands_History ens).

By 1976, with economic surpluses in hand and Syncrude on its way, Lougheed established the Alberta Heritage Savings Trust Fund, ($1.5 billion initial deposit and a promise to put away another 30% of future oil and gas royalties annually). Syncrude shipped its first barrels of oil in 1978, and Lougheed's PCs soared in voter popularity. The boom was on. Frank Spragins, Syncrude's chairman from 1975 to 1978 (starting with the company at the Mildred Lake project in 1965) described the economic impact of the new Syncrude plant (Figs. 4.3 and 4.4), implying an equitable distribution of benefits, with particular focus on the provincial capital.

> The project has provided thousands of jobs for Edmontonians. For almost three years, Syncrude paid an average of approximately $1,600,000 per day to the people of Edmonton for wages, goods and services. This is an average of 3 dollars per day for every man, woman, and child in the metropolitan area. Dozens of engineering firms, pre-assembly and fabrication shops and literally hundreds of goods and services companies were active … between 10 and 11 thousands people directly employed on the Syncrude site. (Spragins et al. 1978).

Fig. 4.3 Glenbow Archives. NA-5059-1 Frank K. Spragins viewing a dragline at work at the Syncrude mine, Fort McMurray area, AB, July 1977

Fig. 4.4 Glenbow Archives. NA-5059-6. Syncrude plant, Fort McMurray area, AB.[ca. late 1970s]

Fig. 4.5 PAA 1190/5 Ft. McMurray Trailer Park 28 Feb 1974 courtesy of the Provincial Archives of Alberta

Fort McMurray grew from a frontier population of 2,000 in 1964 when Great Canadian Oil Sands (Suncor) began work, to 9,542 people by 1974. By 1976, the Syncrude workforce rose to 10,000 and by 1978 the population had reached 35,000 (Marsh 2005:656). Jobs were plentiful, and experience unnecessary for many. Housing was hard to find, though; workers slept in tents in the summer, or makeshift trailers. Others rented rooms in apartments and basements or commuted. For the mostly young workforce, Fort McMurray was a good place to make money and dream big. Wages were high, even for the low-skilled or noncredentialed. One former boom-times worker, who went to Fort Mac "for 4 days and stayed 7 years," described the community as "bush-cosmopolitian," capturing the casual optimism of the first boom. He and his wife found a city full of Canadian and international professionals, tradesmen, and unskilled workers attracted to the employment, high wages and lifestyle of a northern frontier town with easy access to boreal wilderness, outdoor recreation, and more. After long shifts of work, the forests, lakes, and good wages provided a means to recharge (Interview with D. West, 2010) (Fig. 4.5).

The original Syncrude project created an influx of workers for industrial, home, and commercial building construction, as well as service work in shops, restaurants, and bars. For the lucky few, the boom meant catching on with Suncor or Syncrude, and moving into new housing in brand new subdivisions in Fort McMurray with company mortgage assistance. For most others, it meant months of shared or temporary housing, or hot bunking in work camps, as the inflated price and limited availability of housing became out of touch, especially for those who

Fig. 4.6 PAA 1190/7 Fort McMurray Views 28 Feb 1974 courtesy of the Provincial Archives of Alberta

provided services for the highly paid industry workers. Economic growth was enjoyed by Edmonton and Calgary as well, as they provided manufacturing, fabrication, engineering, and business services to the tar sands. Alberta experienced unemployment rates between 3.7 and 4.7% from 1975 to 1981, and the cities of Calgary and Edmonton grew by 66 and 58%, respectively (Krahn and Harrison 1992) (Fig. 4.6).

The Bust

In 1980, with the price of oil again on the rise, Liberal Prime Minister Pierre Trudeau announced the National Energy Program (NEP). Intended to secure prices of oil and gas for Canadians, the program heightened political tensions between Ottawa and Alberta. The problems worsened when Ottawa refused Syncrude and other oil companies the right to sell at world oil prices. Widely perceived as a financial rip-off by Albertans at the time, researchers estimate the NEP siphoned between $50 to $100 billion from Alberta's coffers (Vicente 2005; Mansell and Schlenker 1995) (Fig. 4.7).

The relationship between the NEP and the bust was immediately drawn by most Albertans. The recession hit Alberta in late 1982 and hardened as oil prices bottomed out in 1985. Oil land leases plummeted. According to Krahn and Harrison

Fig. 4.7 Glenbow Archives M-8000-581. Cartoon by Innes, Tom, Calgary, AB. Calgary Herald. 1 April 1980. Accession comments: "The federal government cancels a clause in taxation and pricing agreements with Suncor and Syncrude which would have allowed them to receive world prices for their products. Lalonde, Minister of Energy, Mines and Resources, deflating the tar sands projects"

(1992:196), "what sets Alberta apart is that the shift from boom to bust was more sudden and dramatic than elsewhere in the country, [and] the boom years of 1974 to 1981 constituted a bitterly remembered counterpoint to the economic deprivation that followed." Using Statistics Canada reports, they discovered that unemployment rates increased from 3.8% in 1981 to 7.7% in 1982, to 10.6% in 1983, and peaked at 11.1% in 1984, before dropping to 8.8% in 1988. People began to leave the province in search of work and life options elsewhere. While Alberta experienced a population growth of over 265,000 people from 1974 to 1982, in the next 6 years the province experienced a loss of almost 120,000 (Krahn and Harrison 1992). Edmonton and Calgary suffered a dramatic drop in national income standings (Calgary from 3rd to 18th and Edmonton plummeted from 5th to 46th), as the poverty rolls burgeoned.

The bust hurt many ordinary working people, and a number of small businesses went bankrupt. However, drawing on surveys and interviews with Albertans, Krahn and Harrison (1992:204) found that the large urban middle classes took the bust particularly hard, forming collective interpretive frames of the "boom to bust" crash that emphasized their relative deprivation. These frames defined the sudden unemployment, loss of income, inability to make debt payments, and changing living conditions as caused by economic structures and swings, not by personal failures. Twenty-five years after writing *Prairie Capitalism*, Larry Pratt mused:

> [I]n 1986 the price of oil collapsed to nothing; it fell to $9. Ten thousand scientists, geo-scientists, engineers, you name it, were fired in Calgary. Instead of looking for other jobs, a lot of them pooled their resources and set up companies, and from those companies come these big [Canadian] ones [like Encana]. It's kind of a Schumpeterian creative destruction … (Mouat 2005).

The pull out of major US and international oil companies from Alberta may have made space for Alberta's new oilfield entrepreneurs, many of whom would lead the next tar sands boom at the turn of the century. But in the mid 1980s that future was not on the imagined horizon of Albertans. Analysts describe how the middle class's experience of a rapid bust was severe enough to alter political allegiances. Unionism increased. Suncor workers that managed to keep their jobs engaged in a bitter and still-remembered strike for 6 months in 1986 that divided friendships and families (Boychuk 1996). Lougheed's popularity declined (Lizée 2010), and support for the left wing New Democratic Party reached its historic peak in the province near the end of the decade (Nikiforuk et al. 1987; Laxer and Harrison 1995). After steadying at 30,000 people in 1981, Fort McMurray's population rose slightly to about 34,700 by 1991, then dropped to 33,900 in 1996. A specialist described McMurray as "a town of uncertainty" whose future could either quickly go up or plummet given the precariousness of boom-bust cycles (Bone 1998:252). The economic rollercoaster in the tar sands left a psychological legacy for people in places like Fort McMurray, where even in the good times of the last decade, "the local citizens are wary of the term boom, coming as it does with its inevitable twin: bust … . Fort McMurray may be the last boomtown" (Gillmor 2005).

King Ralph Launches a New Era (and Shreds the Public Interest)

Syncrude was crippled by increasing costs of production and lower oil prices of the mid 1980s. The oil sands companies renegotiated royalties downwards in 1986, but it would be a long struggle for Syncrude to reach a scale of production that generated profits, and even longer to generate royalties for Albertans given their generous capital write-downs. Nevertheless, by the early 1990s, the Syncrude and later Suncor plants must have appeared to many Albertans like pinnacles of recovery. The politician who would see this apparition to reality was just emerging.

Ralph Klein, a former journalist, mayor of Calgary (the financial and management headquarters of the oil industry), moved from his position as Environment Minister in the government of oilman Don Getty (1985–1992), to take the helm as the new Conservative Premier in 1992 and remained there for 14 years. He was deeply familiar with the problems faced by the industry during the recession and would prove a formidable proponent of oil and gas development, and the corporations that enabled it. Ralph's folksy, straight-talking manner made him a winning media personality. Skilled at managing criticism, he overcame public protests to his deep cuts in government services, and defended new tax and royalty breaks for pulp and paper, forestry, and petroleum industries as necessary for economic diversification (Taft 1997).

Clarke (2008) describes the Canadian political surrender that ended the experiment with energy nationalism under Trudeau, and the introduction of the 1987 Free Trade Agreement between Canada and the United States which sealed nationalism's fate: "The FTA meant that Canada's energy future would be driven by free market energy policies' that put control in the hands of multinationals, and limited the role of Canadian federal government over tar sands" (2008:47). In the early 1990s, Klein and the Federal government worked with the industry to put in place a royalty regime designed to encourage the growth of tar sands mining and diversification of capital investments into refining and upgrading in the province. Sweeney describes the impact of the 1995 National Oil Sands Task Force Report (2010:130):

> The Task Force outlined a twenty-five year growth strategy for the Sands, calling them 'the largest potential private sector investment opportunity for the public good remaining in Western Canada, and 'a national treasure'. The paper proposed investing up to $25 billion to boost production in stages from 450,000 barrels a day to nearly a million. The task force confidently predicted that all of this activity would create 10,000 direct new jobs. ... the Alberta government was asked to consider an across the board 1 percent gross royalty until companies could pay off their multibillion-dollar investments. The Task Force also asked for a reduction from 50 percent to 25 percent on profitable production after recovery of capital costs.

The authors urged reductions in federal corporate taxes, and described bitumen as a "knowledge-based, technology-driven, resource of substantial quality and value." It would take another decade for the rhetoric about the knowledge economy to work its way into the government vernacular, but Klein put the proposed new royalty regime in place by late 1995 (Plourde 2009). Canadian journalist Don Gillmor (2005) described the additional support from the Federal Government:

> The next boom came in the unlikely form of federal aid from a Liberal government. Jean Chrétien, who had been Trudeau's lieutenant during the National Energy Program era and who might have been lynched had he come to Fort McMurray in 1980, arrived in 1996 to publicize extensive federal tax breaks for oil sands development. Companies could write off 100 percent of their capital costs, including overruns, in the year they were incurred. The provincial government introduced an incentive package that charged only a 1-percent royalty on oil sands revenues until capital costs were paid off. The effect was immediate: housing starts in Fort McMurray rose almost 300 percent within twelve months, and the leases along the Athabasca River quickly solidified into a checkerboard of international interests.

Acknowledgement of the public ownership of the resource was lost in the cacophony. The Federal Government began to privatize the government-controlled share of Petro-Canada in the 1990s, including its share in Syncrude; Alberta sold their 50% share in Alberta Energy Company in 1993, also part of the Syncrude consortium. The North American Free Trade Agreement of 1993 linked future Albertan tar sands export levels to the US economy (and US consumers) in ways that would further limit political options for Canadians (Marsden 2007). Premier Klein urged government to "get out of the business of business," arguing that only privatization and the free market could develop public resources effectively. Klein's symbolically loaded discourse justified his new 1% oil sands royalty rate, among the world's lowest, combined with deferred taxes and royalties until project expansions were completed, as astute incentives (and not subsidies) that would enhance the public good (Sweeney 130). Meanwhile, as Albertans soaked up these symbolic accolades, privatization directed massive tangible benefits to the corporate few. Whenever faced with criticism from Ottawa regarding Alberta's weak environmental regulations, "King Ralph" would remind Albertans of how, as Mayor of Calgary, it pained him to see Westerners suffer because of Ottawa's 1980s National Energy Program. Deftly turning the tables, his rhetoric linked environmental regulation to negative economic outcomes, focusing public resentment back against his critics. Alberta politicians continue to draw on this master trope to shut down public dialogue. Similar verbal imaginaries are used to bludgeon critics of the current boom, and its social or environmental impacts. Corporations are portrayed as victims of Ottawa's poorly thought out global warming policies and the international interests embodied in the Kyoto Protocol. Recollections of the National Energy Program are particularly acute, however, washing away corporate power and influence from popular purview.

In the mid 1990s, technological shifts in the mining operations from bucket-wheel to large truck and electric shovels, and improved processing (decreasing the amount of overburden in the bitumen mix) reduced the input costs per barrel of oil. Sweeney reports that over 400 jobs were lost in the switch, but that the scale of mining increased. Large shovels reached 100 t, and the familiar trucks changed from 60 t in the late 1970s to 240 t by early 1990s, 340 t by late 1990s, and between 385 and 400 t since 2000 (400 t is the approximate curb weight of 170 Ford F150 pickup trucks).

The increasing scale and cost effectiveness of mining had a major impact on profitability. But most of the bitumen remained too deep to mine, a challenge that reinforces the "materiality of nature" in the economics of production. In situ and SAGD were the answers offered by technology – called huff and puff by some as they inject steam into the tar sands deposit to liquefy the bitumen – emerging as Alberta innovations initially in the Cold Lake region. Changes in mining and extraction also began to alter the scale and types of ecological impacts. Meanwhile, in the war of words and images, truck and shovel technologies quickly became the new cultural symbols of progress, familiar icons of progress among Alberta citizens.

The Current Boom

> The province [Alberta] can control its own destiny more than any other because, in the years to come, Canada will need Alberta far more than Alberta will need the rest of Canada. … And there you have the tension that simmers between Calgary and Ottawa. Two capitals: one economic, one political. Both eager to exploit the oil sands for their own, incompatible, often contradictory agendas (Maich 2005).

Shell Canada announced the first integrated oil sands project in over 20 years in 1999 (an open pit mine at Muskeg River and the Scotford Upgrader in Fort Saskatchewan). They became operational in 2003 and are designed to produce

155,000 barrels a day. Sinclair (2011) argues that Shell's Muskeg River operations opened the flood gates for a new investment and construction boom in the tar sands, coinciding with all of the usual side effects for communities, although greater in scale than in the first boom. In Fort McMurray, population again blossomed, 35,000 in 1998 to 65,000 in 2008, the regional population breaching 100,000 in 2010. Housing starts cannot keep up with growth and in 2008, home costs in Fort McMurray averaged over $650,000. A bedroom in a shared unit was going for $1,000 a month, two-bedroom condos for $3000, and houses were going for $4000 a month. Another 25,000 workers (not included in the population figures) were living in camps, motels, illegal basement suites, and vehicles.

The boom's impacts were not so easily swept under the rug this time around, posing a legitimacy challenge for the state. One journalist described the hyper-expansion of the tar sands in 2006 as a "Boom Gone Berserk" (Reguly 2006). One of Klein's more infamous parting phrases summed it up in surprisingly frank terms; in response to allegations that the state did not do enough to address the boom's impacts, the out-going Premier stated simply: "We had no plan." Can such state inaction be explained by Alberta's position as a prototypical petro-state? There are some indicators that the petro economy has shaped the Alberta state: a high degree of dependency on energy staples has made Alberta a raw materials price taker; long-term ruling parties have developed close alliances with the petroleum industry; most of the government's revenues derive from oil and gas rents and royalties; the circulation of industry elites to and from politics or state bureaucracy is noticeable; as is influence and control of industry over policy; there are small under-resourced opposition parties are small and under-resourced; and with such low personal taxes paid by citizens, the state appears relieved of its contractual relationship with the public. And while even Lougheed recognized that conventional oil and gas deposits in the province would eventually decline, and planned for the eventual withdrawal of major energy companies, this insight is lost on Lougheed's contemporaries. During the Oil Sands Consultations (Alberta 2010), one member of the public put it this way:

> I recently read an article that made claim that Alberta is on its way to becoming a post democratic state, and I can't say that I disagree with it. I don't feel that we have world-class governance. I feel that we have corporate governance with an unhealthy fixation on the economy. (Edmonton, 4 April 2007).

But there are other indications that state actions are not so readily explained by structural conditions. The state has in fact acted autonomously in several instances, having been actively engaged in promoting the commercialization of the tar sands, and today there are multiple development models from which the state could have chosen. Finally, while perhaps not the majority of the local citizenry, there is a vocal and growing opposition to this "plan-free" approach to development with which the state must contend. What is more, energy-producing states today – even those fitting closely the ideal-typical depiction of a petro-state – must also contend with international pressures other than those imposed by the corporate elite. Alberta's position on climate change and the environmental impacts of the tar sands, climatic and otherwise, have drawn considerable international criticism. Alberta and the petroleum industry have responded to their global critics with vigor (and money) in media and political campaigns, generating a dynamic political dialogue in which contentions for legitimacy are heated, the outcome uncertain. More on this in the next chapter; in what follows here we draw on the discursive record to unravel the competitive meaning-making occurring around these boom impacts for labor, families, and community at home in Alberta, and reflect on the mixed benefits and complex risks faced by Albertans, including First Nations and Métis peoples. Key themes emerging in the discourse are pace of development, labor, and royalties.

How Fast is Fast Enough? The Pace of Development

> The development of Alberta's tar sands has outpaced government policy and planning. The hare has disappeared over the horizon leaving the turtle choking in the dust. (ENGO representative, Oil Sands Consultations, Calgary, 23 April 2007).

Despite Klein's admission of no plan, there were in fact efforts made to coordinate, if not control, tar sands development. With so many construction start ups, in October 2005 the Alberta Government announced its Mineable Oil Sands Strategy (MOSS), which essentially consisted of zoning the landscape to favor particular uses, to replace the aging Fort McMurray Integrated Planning Strategy. Public consultation on the MOSS proposal barely began before it was halted by widespread protest. A new public hearing process, designed by Alberta MLAs, was enacted in late 2006 "to establish a process that allows Albertans to share their vision and principles for development of Alberta's oil sands." (Alberta Environmental Law Centre 2007). Meanwhile, Premier Klein had announced his retirement earlier that summer and the political contest to replace him took off in the late fall of the same year.

Klein's "no plan" comment offered a rallying point for citizens increasingly concerned about tar sands expansion, and the Oil Sands Consultations, launched in Fall 2006 by the Progressive Conservatives in an attempt to temper swelling local opposition, would have the opposite effect, becoming a lightning rod for citizens, particularly as the symbolic sentiments offered during the hearings contrasted starkly with the events taking place beyond the hearing rooms. Testifiers focused first and foremost on the rapid pace of development and the lack of any comprehensive public input into decisions until after leases were signed and sealed (Way 2010). Public discontent flared higher when transnational oil companies announced plans for upgrading and refining in the United States, instead of Alberta. The public felt that government had betrayed their dream of value-added industries as jobs went down the pipeline to refineries in the United States. Clarke (2008: see maps 91, 145) argues that the pipeline deals established a web of infrastructure that rendered Canada an energy satellite.

One sure sign of a fragile state legitimacy, or perhaps an ideology run amuck, is the emergence of critique from unexpected social locations. One such sign came in the shoes of former Alberta Premier Peter Lougheed, whose entrance was described by Globe and Mail reporter John Gray as an emergence from a 20-year hibernation. Lougheed took aim at NAFTA, among other things, lecturing the US Ambassador and other prominent American officials at a Calgary barbeque on how Albertans "were the owners of the resource" and should be free, in a seller's market, to sell oil to China and other Asian countries, rather than be bound to the Americans. He also targeted Klein, offering an influential voice to the groundswell of concern about plan-free planning: "We should have more orderly development ... That means, do one plant, finish it and build another plant, finish it, do another plant – instead of having four of them going at the same time" (Gray 2008).

According to Lougheed, Klein was responsible for overheating the economy and wage earners outside the oil industry faced unnecessary inflation and cost of living increases, especially in retail and housing markets. Conservative Party hopefuls running to replace Premier Klein as leader distanced themselves from the scuffle, although they were careful to assure voters that they would support oil sands development in general. Originally, Lougheed sympathized with Stelmach, who inherited Klein's mess after all, but a few months later he forecast political disaster for the party because of the new Premier's mishandling of the issues. In August 2007, Lougheed offered a speech to the Canadian Bar Association in Calgary, in which, according to one journalist present:

> Peter Lougheed ... issued a stern warning; we're heading towards a nasty constitutional dust-up that could see Alberta pitted against the rest of the country. The source of the friction? Accelerated tar sands development that is causing serious distress to the environment. Resource development is a provincial matter, while environmental policies fall mainly under federal jurisdiction. At some point, and it could be sooner rather than later, said Lougheed, one level of government is going to have to take a stand against the other (Stewart 2007).

Ordinary Albertans began quoting Lougheed in their own presentations at the consultation hearings to support claims for a moratorium on oil sands development. Media reports also frequently cited Lougheed's suggestion that industry pays half the costs of twinning the 212 km of Highway 63 from Fort McMurray towards Edmonton (then at a projected cost of $680 million).[1] Others reported Lougheed's bitter criticism of government acceptance of industry plans for processing Alberta bitumen in United States' refineries: "These are our jobs. The best jobs in the oil sands are on the refining side of it. We should be having the upgraders here" (Makin 2007; see also Gray 2008).

The Oil Sands Consultations were held throughout the Athabasca Tar Sands region in Fort McMurray, Wabasca, Bonnyville, and Peace River, downstream in Fort Chipewyan, and in the southern cities of Edmonton and Calgary, from September 2006 through April 2007. Critics repeatedly proposed a slowdown to let government, municipalities, the economy, and the public catch their breath. At the hearings, many members of the public found Lougheed's position to have more validity than that of Premier Stelmach. But Lougheed was not the only unexpected, and influential, voice of concern. Even Mayor Blake of the Regional Municipality of Wood Buffalo (Fort McMurray) said "enough is enough" and she and others urged planning for infrastructure deficits, housing shortages, and negative social impacts of the boom.

---

[1]For cost on highway see *Canada and Alberta partner to twin Highway 63,* available at: http://www.infc.gc.ca/media/news-nouvelles/csif-fcis/2006/20060829fortmcmurray-eng.html. Accessed 20 Dec 2010.

Many expressed frustration with government refusals to complete social, ecological, and health impact studies first, before approving the tar sands leases:

> Our premier, Mr. Stelmach, said, oh, heavens, if we don't continue with the fast pace of development of the Fort McMurray tar sands and the Athabasca tar sands, we're going to lose all kinds of business. Where's the oil and gas going to go, ladies and gentlemen? (Oil Sands Consultations, Edmonton, 3 April 2007).

> I hear often of people living homeless, many of whom have jobs, while others simply can no longer afford to live with the rising costs of living. Studies to measure the impacts continue, but meanwhile, projects keep getting approval. I'm told, however, not to worry, because the market is supposed to know what's good for our communities, for our people, or for our neighbors, but I'll ask each and every one of you to tell me how this market, who's [sic] only determination of success is profit, is going to ensure that no more people will be left homeless, that oil sands will not be our single largest contributor to greenhouse gas emissions, that no more children will be kept indoors at recess, or that no more people will die. (Oil Sands Consultations, Edmonton, 3 April 2007).

The previous speaker justifies a slowdown because the value of the resource will only increase as reserves decline, and he goes on to ask whether the idea of success in the free market includes social justice, basic needs, housing and jobs, or simply making corporate profits. Members of the opposition parties in the Alberta Legislature agreed:

> Today I would like to speak about the urgent need for planned, orderly growth in our province, particularly in the tar sands. The current gold rush style of growth is leaving many Albertans behind. Families are feeling the squeeze. There aren't enough schools. Wait times in hospitals and emergency rooms are increasing. Families are facing high tuition fees for their kids and a shortage of affordable, high-quality long-term care for their parents. Working families face increasing housing costs, homelessness in their communities is increasing, and there is serious environmental damage. Despite the rhetoric in yesterday's Speech from the Throne it's clear that when it comes to planning, this government just doesn't get it. (Mason. Alberta Legislative Assembly, 8 March 2007).

Premier Stelmach never attended the Oil Sands Consultations. But his speeches in other venues offered in the same time frame nonetheless "spoke" to testifiers. Speeches offered to professional and business groups take the form of disciplining the masses:

> To halt growth would be to deny our ability to meet growth's challenges with Alberta's leadership, innovation and entrepreneurial spirit. (Stelmach. Canadian Energy Research Institute Oil Conference, Calgary, 23 April 2007).

> While we're on the topic of the oil sands, I want to make one thing very clear. There has been talk by the federal Liberals—and others—of a moratorium on development of Alberta's oil sands. My government does not believe in interfering in the free marketplace. You cannot just step in and lower the boom on development and growth — in the oil sands or elsewhere. If that were to happen, the economic consequences for Alberta, and for the economy of Canada, would be devastating. (Stelmach. Rotary Club of Calgary Valentine's Luncheon Address, Calgary, 13 Feb 2007).

To commit such a rash act as controlling the pace of growth, he argues, would have devastating consequences:

> It's no secret that Alberta's tremendous economic success is due in large part to the energy industry. Alberta today is indeed an economic powerhouse ... Alberta is the Canadian leader in almost all economic indicators, from income and job growth, to education of the workforce and standard of living ... . My government is committed to ensuring there will never again be a major downturn like we saw in the 1980s. A key part of this is our commitment to

developing and diversifying the energy industry, and diversifying our economy as a whole. This includes encouraging more upgrading and value-added activities here in the province (Stelmach. Canadian Energy Research Institute Oil Conference, 23 April 2007).

Drawing on the legacy of resentment towards Ottawa and the 1980s bust, he hooks deep into the Albertan psyche, and the phrase "lowering the boom on development and growth" then chastises Albertans – including a few of his fellow Conservatives – for suggesting such nonsense. In this storyline, critics become treasonous outsiders and federalists, regardless of just how rooted the critics really are, like long-time residents Lougheed, scientist David Schindler, or Aboriginal community leaders downstream in northern Alberta. A prudent leader would never put on the brakes; the real burden of leadership is to forge ahead – for the benefit of the whole country, not just Alberta.

At the Economics Society of Calgary, the Premier first pumps up his corporate and business listeners with tales of power shifting to the Canadian west from Ottawa, then repeats his warnings:

In Canada, too, economic power is shifting. Alberta's economy continues to lead the country. In fact it's fair to say that Alberta is the engine that drives the Canadian economy and produces prosperity not just for Albertans but for all Canadians. However, the pressures of growth are putting a strain on Alberta's human and physical infrastructure… and on our relations with other Canadian jurisdictions. On that last topic, we see some amazingly short memories among those who speculate on a moratorium on oil sands development. When energy prices and Alberta's economy sank in the early 1980s, the impact was felt across the country. (Stelmach. Economics Society of Calgary Luncheon, Calgary, 22 Feb 2007).

As Alberta's economy goes, so too does the country, is his thinly veiled threat to the nation. But across town, at a public meeting of the Oil Sands Consultation Board, this citizen wondered whose interests the Premier represented:

I'm [deleted] years old and in the [deleted] years that I've lived in this province, the one thing it has taught me is not to trust the government, and I think that's really sad because I would love to believe that you guys want to do the right thing and that if you hear that Albertans want a slowdown and that we want to make sure this is done in the way that benefits our people, I want to believe that that's what's going to be what happens, but I don't know. I just don't know if I can trust that. This morning we actually organized a protest over at the Palliser—I'm feisty and I like protests—and basically we protested because Ed Stelmach was over there at the Palliser giving the opening keynote address to this big oil conference that was sponsored by Halliburton and Shell and Suncor, and he was basically saying to them, you know, Hey, let's go, I'm going to talk to the private investors…and I was kind of wondering, well, why aren't you here at the consultations listening to what the people of Alberta have to say because to me that's far more important, and if folks—at least in Edmonton and …In Peace River we're saying, we want to slow down and we are concerned, then he shouldn't be going over there saying, well, business as usual, folks, let's shake some hands, and that really floored me, so we did a little bit of a media kind of event and it felt good, but what would feel a lot better is if we actually saw some real political action and some real political will. (Oil Sands Consultations, Calgary, 23 April 2007).

In Edmonton, members of the political opposition in the legislature raised similar concerns about the inequitable distribution of benefits and risks and called for new resource rent rates, reductions of greenhouse gases, monitoring of cumulative environmental impacts, and preparation to mitigate health and social impacts.

> The Alberta advantage is becoming a disadvantage for more and more and more people. …
> That's why over here on this side of the House we've been saying that we have to slow down
> this development. Nobody is really benefiting other than a few CEOs in downtown Calgary.
> We've talked about rents, we've talked about health care not being able to keep up, and at the
> same time we're heading towards an environmental disaster with more and more CO_2 being
> there, Mr. Speaker. That's why we've said: "Look. For the time being let's at least slow it
> down, have a moratorium, figure out what kind of Alberta we want in the future, figure out
> what makes sense in terms of how much CO_2 we can keep putting out, see what makes sense
> in terms of our needed social programs." (Martin. Legislative Assembly, 17 April 2007).

The Premier resisted such calls. With a year of experience under his belt,
Stelmach stridently claimed that a slowdown would be un-Albertan, and a moratorium would deny Albertans their heritage:

> Albertans have always been innovators. We've had to be, right from our province's first days.
> Where others saw a remote frontier, Albertans saw opportunity. The place some people
> thought of as an isolated outpost, too far from the established centers of Ontario and Quebec,
> the United States, and Europe now boasts its own style, as the home of great innovators,
> global companies, world-class universities, and cultural leaders. When others wrote off the oil
> sands as too hard to develop, too expensive to produce, and too far from markets, Albertans
> showed great determination in developing this resource. They found ways to make it work…
> and a multi-billion-dollar industry has been created… One that promises prosperity for our
> province, and energy security for Canada and North America as a whole. It seems to be in
> Albertans' DNA to question accepted wisdom… to look for better, more efficient, and smarter
> ways of doing things. (Stelmach. Jasper Innovation Forum, Jasper, AB, 12 June 2008).

It is thus not the arguments posed by critics, but their very DNA that is subject to
question and hence marginalized from what should rightly be a discussion only
among true Albertans. The Premier weaves a series of metaphors and storylines to
negotiate consent for his full steam ahead approach, tapping into the pride, ingenuity and innovative spirit of Albertans. His assurances that together we can solve any
problems that come our way offer confidence that the current malaise is merely a
wrinkle to be worked out.

The Faces in the Machine: Working People

Remoteness, the defining spatial feature of the Athabasca region, poses constraints
to capital. Making oil from bituminous sands requires an elaborate suite of production conditions, and the further along the spectrum of nonconventional forms of
fossil fuel, the greater the labor and mechanical requirements to transform raw
materials into commodities. Labor (professional, skilled and unskilled), equipment
(mining, extraction, coking, refining, transport), food and shelter, all must be transported to a remote region 500 km north of Edmonton. Mining and industrial-scale
processing operations have to function for half the year in extreme cold, snow and
ice and, another quarter of the year in mud and bog.

The city of Fort McMurray serves as permanent home to many tar sands employees, who come from all over Canada and around the world, but an even greater
proportion of the workforce is mobile, working in one place, living in another.

Fig. 4.8 Camp N 57.11.16 W 111.34.59, Albian Sands Muskeg River Mine, AB, Canada with permission of Louis Helbig. See www.beautifuldestruction.com and http://www.louishelbig.com/

Estimates of the size of the mobile workforce vary between 16,000 and 25,000. Half come from Alberta and half from outside the province. Most are males. Three quarters stay in camps (Fig. 4.8).

Many maintain a primary residence in other urban locales outside Ft McMurray and commute between shifts of 10–14 days (or more), 4–6 days off. Shutdown workers – those engaged during interruptions in operations – and those on special construction projects with tight timelines often work straight through until a project is complete, sometimes 20 days at a time. Some workers rent temporary residences in Fort McMurray. Workers in camps or in town commute, by individual vehicle or chartered bus, back home to families and partners in Edmonton and other southern communities. A few fly back to homes further afield. All day, every day, Diversified Transport runs about 500 buses full of workers back and forth from Fort McMurray and the work camps north of the city to the construction and plant sites, as well as coaches from Fort McMurray to Edmonton. Their brochure claims that the company's buses travel "around the globe 1.8 times" every day. Likewise, annual airline passenger movement to the city has increased from 223,000 in 2004 to 704,000 in 2009 with more than 195 flights a day (Fig. 4.9).[2]

Newfoundland provides the largest supply of people who make the "Big Commute" (Storey 2010): of these about 40% are regular commuters who fly back and forth between shifts, and 60% stay for 6 months or longer as temporary seasonal workers.

---

[2] Information available at: http://www.flyfortmac.ca/AirportAuthority/Expansion.aspx. Accessed 9 Dec 2010.

Fig. 4.9 Buses on way to
work sites B2402276 with
permission of Louis Helbig.
See http://www.
beauifuldestruction.com and
http://www.louishelbig.com/

Many Newfoundlanders came West following the closure of resource industries in their home towns (ibid.), including many fisherman since the collapse of the cod industry, and former employees of the recently closed Abitibi-Bowater Pulp Paper Mill at Grand Falls. Newfoundlanders also came to the aid of the fledgling industry back in 1976, when the *Come By Chance* oil refinery in Labrador closed, some of whom make up Ft Mac's permanent residents today (Storey, 2010).

The experiences of this increasingly migratory staples workforce should set off warning bells for Albertans reliant on tar sands jobs. Unless communities, governments, and unions plan otherwise, when the natural resource is exhausted, corporations do not create new jobs to replace them, and workers generally do not get "retrained." Instead, they become subject to the geographic whims of the global primary industry labor flow. Unfortunately, when the Alberta energy boom subsides, and if no alternate energy sources emerge, many of the employment fixes available to former extractive workers who lost jobs in one region in the 1970s and 2000s may no longer be there, if there is no more energy to fuel another round of extraction.

In Alberta, mobile workers are classified by planners as shadow workers who do not reside in a region or pay municipal taxes, but nonetheless place heavy demands on municipally funded infrastructure like water, sewer, roads, recreation, policing, social, and health programs. For example, trades people may work a series of planned shutdowns for different oil sands companies, staying in camps until each project is complete, then return home, and after a break move to the next shutdown. Like seasonal workers, mobile workers tend not to develop a sense of place that is so crucial for motivating personal investments in community and civic responsibility, and cumulatively this lowers community resilience. In a sense, we have two Fort Macs – one desperately trying to assert a sense of community autonomy and identity, and the other full of people just there to extract paychecks.

Not all workers, moreover, enjoy the fruits of high paychecks, or even jobs. Although the opportunities would appear limitless in Fort McMurray, the naïve job seeker learns otherwise. Recruitment and hiring processes for the operations and mining jobs are much more formal than in the first boom. It may take weeks to schedule interviews with major oil companies, or for reference and drug checks to clear.

> We have people that come to our communities every day. Some of them are from communities like [Fort]McKay, Janvier, Conklin, [Fort] Chip[ewyan], and some of them are from all over the world, and they show up in Fort McMurray, some of them because they have no place to live anywhere else. That's true. Others because they do have some place to live, someplace else, but they have come to the land of milk and honey where the streets are paved with gold and you've all heard the story, and they come and knock on my door or [indiscernible] or many of our doors maybe, and say, great, where do I go to get the job and get put in camp. And, sorry, and next thing you know somebody who has never been homeless before a day in their life is homeless in Fort McMurray (Oil Sands Consultations, Ft McKay, 27 March 2007).

The workforce in the oil sands sector is predominately male, describing the second key source of stratification in the tar sands labor force. Few women are hired and those who are, report more than the usual discrimination, paternalism, racism, and barriers to traditionally male occupations (O'Shaughnessy and Krogman 2010). For the many mobile workers, camp life also offers highly divergent experiences, with each company ranging from first class camps with well planned layouts, good food, and recreation options, to poor quality camps with few services and poor worker morale. All are near worksites, and some workers feel proximity leads to overwork and stress. Others report that camps "feel like jail" with tiny rooms, routinized schedules, security guards babysitting workers, swipe in and swipe out privileges to eat or to leave the camp, lineups for food, and sleepless nights as shifts change. The separation from family and networks of friends at home for weeks on end also takes an emotional toll, on the worker and on the family unit. Some report a lack of freedom to raise concerns about poor food, open abuse of stimulants, lax upkeep of facilities, or unsafe working conditions (Dembecki 2010; Angell 2010). Recently, one tradesman chose to blog about his negative camp experiences, poor food and camp upkeep, and the poverty of emotional and social life inside the camp. He was fired, allegedly for posting photographs and video of inside conditions. Others describe reluctance to report injuries on the job. Such pressures create a culture of silence, especially in trades where the next contract depends on being recalled.[3]

---

[3] http://adhdcanuck.wordpress.com/2010/08/23/work-camp-life-can-range-from-excellent-to-rotten-allow-me-to-introduce-you-to-rotten-enter-the-mackenzie-camp-zone/ Accessed 29 Nov 2010.

Fig. 4.10 Ed Stelmach is soiled by magazine's Oilsands coverage. Malcolm Mayes (2009), Image MAY2261 with permission of ARTIZANS and Malcolm Mayes. Alberta's Premier reads the National Geographic special issue "The Canadian Oil Boom: Scraping Bottom." The article and photographer Peter Essick's images were widely reported in major Canadian newspapers and television news shows: the "baby seal moment for Alberta's oilsands." http://news.auroraphotos. com/2009/03/peter-essick-images-of-oil-boom-receive-press/. Read the National Geographic article here (http://ngm.nationalgeographic.com/2009/03/canadian-oil-sands/kunzig-text) and see Peter Essick's full set of oilsands images (and pithy captions) at Aurora Photos: http://www. auroraphotos.com/SwishSearch?Keywords=essick+mcmurray&submit=Go!

Downtown McMurray is the social outlet for many workers after the whistle. The magazines *Chatelaine, National Geographic*, and the UK newspaper *The Guardian* have each published graphic assessments of the darker side of life in Fort Mac for workers and citizens. Bars are full. The Boomtown Casino, one of the largest money-makers in the province, is packed. Health officials report the highest rates of syphilis and STIs in the country. Taxis charge top dollar to pick up and deliver people and booze back and forth to the camps an hour or more north of the city. *Chatelaine* depicted women as gold-diggers, strippers and escorts, and highlighted abuse of drugs – reflected in popular nicknames like Fort McCrack and Fort Meth. Other stories describe the homeless and the working poor who cannot make ends meet in the City despite earning $70,000 a year, or depict a masculine culture of blue collar workers, who measure their manhood by the number of dangerous jobs, overtime hours, dollars earned, or toys purchased. In 2009, *National Geographic* particularly angered long time McMurray residents, who felt that "drive-by writers" only sought to titillate readers with the seedy side of the boom, and failed to capture the resilience and strength of the community (Preville 2006; Edemariam 2007; Kunzig 2009) (Fig. 4.10).

And then There Is the Drive

This road is gonna be the death of me ...
somewhere north of nowhere, blacktop's all I see
... this northern town is keeping my dreams alive,
It's a little bit of heaven ... but the devil owns the drive ...

"Highway 63." Lyrics and Music by Ken Flaherty. K Flat Music 2009.
The long single lane highway servicing the region since the 1970s remains the main route – north – although it takes less time nowadays as trucks are faster and more powerful, road surfaces better, and speed limits higher. But the volume of traffic from the south is huge, with the highest tonnage carried per kilometer in the country and some trucks with more than 300 wheels. Rail lines end south of the city which means all equipment is trucked by Hwy 63 through the city to industrial sites north of McMurray. Over 50,000 vehicles cross the Athabasca River on this highway each day. On the long 200 km stretch beginning at the Hwy 55 intersection, driving above the speed limit is a thing of lore among commuters. Winter weather, black ice, and fog and smoke from forest fires create road and visibility hazards. Often 63 is closed or backed up to accommodate oversized rigs hauling trucks, giant equipment, or building modules to the mining and refinery sites. Some days, vehicular accidents – caused by high traffic volume, drowsiness, dangerous driving, and driving under the influence of drugs and alcohol – require closing the highway. Among drivers, the road's nickname is "the highway of death." The Royal Canadian Mounted Police (RCMP) reported 23 deaths and 22 serious injuries in 2007. A tragic accident near Christmas 2008 took the lives of four Filipino workers, three of whom were temporary foreign workers on their way to Edmonton.[4] For weeks in 2010, reaches of the road remained without emergency services because local volunteer firefighters withdrew support, too psychologically drained, from tending at least once a week to the seriously injured or killed.

---

[4]http://filipinojournal.com/v2/index.php?pagetype=read&article_num=01092009 _issue = V23-N1. Accessed 29 Nov 2010.

In recent speeches, the Mayor of Wood Buffalo and other officials describe another side of Fort McMurray, as a place to raise a family, build a life and career. These spokespeople strive to distance themselves and their town from the images that arose during the boom, and the "work hard, play hard" motto. A recent news article captured Mayor Blake's sentiments: "That's what we're trying to guard against, to make sure that whatever pace we do end up following is one we can sustain." Many tar sands employees are making homes in Fort McMurray and, the Mayor reminds us, there is a core community full of good stories. But this insistent portrayal of social cohesion in the city appears to the outsider as a defensive effort to maintain control over a rapidly disintegrating community identity (Weber 2010).

Big Wages, Little Benefit

Maurice- Qu'est-ce que vous faites ici?
Jaypee- On est venus faire du cash! On est venus faire le gros motton!
From the play Fort Mac by Marc Prescott (2009).

What about those big paychecks? For some in the sands, wages can range from $150,000 to $200,000 a year with overtime. Yet in a 2003 study, researchers at the AFL found that making lots of money is not always a good thing. In their report, researchers identified a series of consequences of the rapid pace of development. Noting that "Canada hasn't seen anything like this since the days of the Klondike gold rush," ... in the "biggest boom since the early 1980s," Even with adjustments, "the Alberta economy has been growing steadily for the last 14 years, with an average annual real growth rate of 3.9%." The authors cite a Statistics Canada study by Cross and Bowlby (2006) showing that "Alberta is in the midst of the strongest period of economic growth ever recorded by any province in Canada's history," with the 12.7% average annual increase in GDP since 2002 comparing favorably with China's 14.8% (AFL 2003). The AFL analysts also identified the negative impact of inflation on wages. In a follow-up study, they found that the rapid growth had increased the inflation effect, much of it energy sector related: "the inflation rate in our province has been higher than the national average for seven of the last ten years." All in all, these factors "make it harder for workers in Alberta to achieve gains in real (after inflation) wages" (AFL 2007 A-5). In Edmonton, real estate prices increased 200% over the decade, and until the recent downturn there was near-100% occupancy in rental markets which drove rents sky-high. AFL analysts observed labor shortages in all areas of the Alberta economy, posing serious challenges to small and medium-sized businesses that could not afford the high wages, and municipalities and their taxpayers suffered too – Edmonton could not afford snow plow drivers, and a major intersection overpass had cost overruns over 100% (from 130 million in 2007 to $260 million by 2009).

Even for oil sector workers whose wages have risen, real wages – that is, the power to purchase goods – declined between 2002 and 2005. Many citizens began to wonder, what good is a boom after all? In 2006, despite labor shortages, wages did not rise (AFL 2007). When ratios between wage rates and corporate profits are considered, even larger inequities surfaced – one of many revelations conveniently absent from proponents discourse. AFL researchers compared the shares of GDP devoted to wages and to corporate profits, and discovered that the share devoted to labor has actually followed a downward trend. In comparison, the corporate share has continued to grow. Other distortions in the labor market are also disconcerting. AFL analysts found that, since the 1980s, a large percentage of jobs being created in Alberta are for part-time or seasonal workers. Despite the boom, the percentage of all wage earners making less than $20,000/year (2004 constant dollars) has barely moved despite the boom:

> While it is clear that unemployment is low, and that there are shortages in some sectors of the economy, it's also true that what some business and government leaders are upset about is a shortage of people prepared to accept lousy jobs (AFL 2007:13).

For the rest of Alberta's workers, the boom was more pinch than plenty. These speakers at the Oil Sands Consultations describe the impacts of inflation for those Albertans who do not work in the oil patch. Two working worlds emerged, one inside the wage boom and one outside, and argued that the state is not protecting purchasing power or addressing the unequal distributions of goods and benefits to the public:

> I hate to say we're making the poor poorer and richer rich in this province now. There's no way that the normal worker that's working in 7-Eleven in Edmonton is making enough money to keep up with the rate of inflation. They're getting poorer on the backs of the people that are getting richer off the oil sands. You have the common workers in Alberta to consider. (Oil Sands Consultations, Edmonton, 3 April 2007).

> I see in my neighborhood an increasing number of working poor, people who struggle because basic costs are increasing all the time, but they are not part of the increased wealth that's flooding our province right now, and that's a big, big concern to me. (Oil Sands Consultations, Edmonton, 3 April 2007).

> I feel the increased amount of economic activity has in no way benefited the majority of the population in Alberta, but only a select few. Of what value is it to have a higher salary when all the other costs, such as housing and other costs, have gone up at least as much? (Oil Sands Consultations, Edmonton, 4 April 2007).

Media reports of corporate profits simply added salt to the wounds. Many speakers at the hearings argued that a portion of corporate revenue windfalls should be captured for the public via taxation or royalties and used to provide sustainable improvements to the "social wage," that is, the range of public services that benefit all citizens of the province:

> On top of the fact that oil companies and their CEOs are making exorbitant amounts of money, the oil companies are constantly asking for taxpayer aid. For example, in the Fort McMurray area where housing is in serious short supply … why aren't the oil companies providing some kind of subsidy since they're making money hand over fist? Taxpayers are paying for a power line to be fast-tracked to meet the upgrader schedule, and it's going to apparently produce twice as much electricity using coal as the city of Edmonton presently uses. Taxpayers get to pay for that. (Oil Sands Consultations, Edmonton, 3 April 2007).

Many costs of the boom then are borne by workers and ordinary Albertans, while the rewards become concentrated in the hands of a corporate elite. The government has

Fig. 4.11 Cover image from Alberta Federation of Labour Report: Lost Down the Pipeline April 2009

stated intentions in the interests of labor, but actions often contradict sentiment (Gilbert 2010). Since the Lougheed years, for example, the province has pursued a long-standing policy of promoting value-added jobs. This meant establishing refineries and heavy oil upgraders in Alberta.[5] The current leadership, however, has signed a series of pipeline agreements that ensure tar sands oil will flow directly to the USA, and be upgraded and refined there by American workers (AFL 2007). More recently, United States refineries increased their capacity to accept tar sands oil, and new United States-based refineries are being proposed. According to the AFL: "What's at stake are literally tens of thousands of short- and long-term jobs, which, without upgraders in Alberta, will be shipped down pipelines to refining facilities in the American Mid-West and Gulf Coast" (AFL, 2009a, b). The very infrastructure for some projects is now made in modules overseas and shipped, railed, and trucked to Ft Mac, such as those for the $8 billion Kearl project that were constructed in South Korea (Fig. 4.11).

---

[5] In response to public pressure, in February 2011 Alberta announced that a new refinery will be built by 2014 in partnership with industry and paid for by processing crown owned bitumen. http://edmonton.ctv.ca/servlet/an/local/CTVNews/20110216/EDM_province_110216/20110216/?hub=EdmontonHome

Contentions over such value-added industries are more complicated than they appear, however. As highlighted by this testimonial, such jobs would come with a price, but one that, nonetheless, we should be paying here, at the mouth of the mine:

> Some might say it would be easier to dig up our bitumen, ship it out, and let some other jurisdiction worry about the CO2 emissions that come from upgrading and refining. But aside from all the potential jobs that would be sent down the pipeline, I don't believe in passing the buck. We could send our bitumen down to the United States Gulf coast where it could be processed by plants that are in some cases 40 or 50 years old and where the regulations might be more lax or we could build a bigger, better, and environmentally cutting-edge downstream petroleum industry right here. (Oil Sands Consultations, Calgary, April 24).

Why does the Alberta government not side with Albertans to guarantee local jobs and protect workers' livelihoods? Certainly the neoliberal ideology so steadfastly embraced by the Tories discourages such interventions in the marketplace. But something even deeper appears to be at play. Many public speakers point to a culture of fear in Alberta that has served to repress protest: fear that they let the last boom slip away in the 1980s; fear that arouses resistance to all manner of suggested intervention, including federal environmental regulations or international standards like the Kyoto Protocol; fear that corporations will pull out if governments ask for higher royalties; fear that investing in value-added jobs might just mess with the program.

The recent global downturn has delayed or slowed some new projects. But, there is little heed of warnings offered by the AFL (2006), among others, that booms inevitably end, and construction jobs usually go first. This warning was also conveyed by a representative from AFL during the Consultations:

> What I would like to draw your attention to is the widely distributed graph for projected construction labor force demand in Alberta. Most of you have probably seen it. Some people call it the collapsing pyramid graph. What it shows is that demand for construction workers in Alberta is currently at about 175,000 and it's expected to peak at about 240 or 250,000 in 2009, 2010. Now, that might sound great, lots of jobs, but there are two problems with this picture. First, as a result of literally two decades of underinvestment by employers and governments and trade training, we don't have 240 or 250,000 construction workers. So, as a result, employers are pushing to make up for this shortfall with ever growing numbers of temporary foreign workers. They're also considering other scenarios to get their projects done without Alberta workers. Senanko, for example, has come up with a scheme that would build – see them building an entire oil sands project in Asia and floating it in pieces across the ocean. At the same time and probably more importantly the projections on construction labor force demand show that by 2015 the demand for construction labor will collapse down to less than 60,000 as major projects are completed. So instead of building our projects over a longer period, we're set to build them in a frantic rush. (Oil Sands Consultations, Calgary, 24 April 2007).

The AFL may be right about the vulnerability of construction workers, but temporary foreign workers, in construction trades or otherwise, are without question the most vulnerable. Unprepared for the sudden labor shortage posed by the boom, Alberta and the Canadian government revised Canada's foreign worker program to ensure ready influx of foreign workers to the tar sands, particularly to fill low-paying service jobs. It seems neoliberals can stomach market intervention in certain circumstances, particularly in ways intended to fuel continued economic growth. Alberta now has the largest proportion of temporary foreign workers (TFWs) of any region of Canada.

Original Albertans

The voices of Aboriginal peoples that most often reach the media are voices of environmental and health concern, as we highlight later in Chap. 5. Aboriginal identity is not quite so simple, however. As much as Aboriginals living down-stream have expressed their alarm regarding environmental impacts, many are seeking wage labor, some simply out of personal motivation, but many in response to a rather shrewd reckoning of their livelihood options – the land base that used to support them, after all, has been so completely torn asunder that a subsistence lifestyle is no longer realistic. The ethical dilemma for many is of course the fact that the very companies responsible for such degradation are the same companies that now serve as the only employment option in the region, but this option may seem more palatable than living off government welfare, or migrating to the cities for work. Employment opportunities are desperately needed if these families are to be able to remain living in their ancestral territories. According to a Regional Municipality of Wood Buffalo Census of 2006, Aboriginal people comprise 12.3% (6,465) of the population. This compares with 5.8% in Alberta. The population is young, about 44% under the age of 24 (Taylor et al. 2009). There are 17 First Nations (16,000 people) and six Métis settlements (6,000 people) in the oil sands region and according to the source cited above, "thousands more live off-reserve and off settlement" in the region. The unemployment rate in these communities ranges between 11 and 33% (Government of Canada 2006), among the highest in the country.

Government claims about Aboriginal involvement in the industry are bold: "In 2008, more than 1500 Aboriginal people were employed directly by oil sands oper-ations, a 60% increase from 1998 … [And the value of contracts] between Alberta oil sands companies and Aboriginal companies was $575 million."[6] Government web pages are careful to present supportive evidence on employment, while acknowledging the "fears of some" Aboriginal people about health and environ-mental impacts. Viewers are offered the impression that government and industry alike have the interests of Aboriginal people in mind, consulting on issues of land management and traditional knowledge data collection, and offering much-needed training and employment.

Research indicates that some Aboriginal people have found oil sands employ-ment and business success. In the case of the Mikisew Cree, much of that success followed a treaty land entitlement settlement signed in 1986 that opened up

---

[6] Available at: www.oilsands.alberta.ca. Accessed 22 Dec 2010.

partnership opportunities with the oil sands industry and its contractors. The stories are promising, but researchers acknowledge that the patterns are uneven and have left certain groups out (Slowey 2008), as this speaker confirms:

> They [Aboriginal people] do not want a handout, and we want a hand up. We would like to know how do we work and how do we get involved with the oil sands industry. I have said the last time that we have men that are gravel truck haulers, I brought one with me last time, and he's a taxpayer. He's also a man that owns his own home, and he pays taxes. He also has his own business. He also is a property taxpayer. In front of his yard he sees gravel trucks going by while his sit in the yard. He's got one sitting in the yard. They are Manitoba plates on them. Our point is to get in; how do we get in? How do we get in and ask for the jobs that are available? We've tried. We've asked – we've gone through the process many times. We've asked many times how do we get the work that's out there. We've asked Shell Canada. They've told us to go to their contractor. We've gone to their contractor. They've told us go to their subcontractor. (Oil Sands Consultations, Peace River, 16 April 2007).

Rhonda Laboucan, elected councillor official at Woodland Cree, Peace River Country, had this to offer:

> The socio-economic impacts in particular with First Nations I find it really ironic our neighbours at the in-situ plant have been asking us to participate in a socio-economic impact assessment but yet here we are a struggling nation trying to make it month to month to be able to provide the program and services to our nations, to our members, and we have a fully able-bodied industry company that can do the work for that said company but yet they don't give us a stitch of work, and they have been there for over 20 years and we have done everything we can to rebuild the credibility and ability of this company, and we've worked with every other player except for that particular company, and I think that's a disgrace and a shame because they have long-term plans in this area. (Oil Sands Consultations, Peace River, 18 Sept 2006).

Recently, Chief Jim Boucher and the Fort McKay First Nation, a community that opposed the industry in the late 1980s and 1990s, have struggled to construct a new relationship with a growing industry that now envelops their community. According to Chief Boucher: "If it wasn't for the oil sands, we would have no economy. … So to be pragmatic and practical about it, the alternative is to do nothing and sit there and collect welfare or to be part of the economy. And I'd rather be part of the economy than let our people be there collecting welfare" (Weir 2008).

These statements of mixed optimism and fatalism are better understood historically, as the authors of a First Nation and Métis Youth Employment study argue:

> Our historical overview suggests that jobs alone are unlikely to eliminate the inequities experienced by Aboriginal communities—inequities that stem, in large part, from historical relations between governments and Aboriginal people. Further, while dependency and economic underdevelopment are unacceptable to Aboriginal people, increased economic development is also seen as problematic by some community members concerned over the growing environmental impacts, continued loss of cultural ways, and worsening social problems (e.g., lack of housing, homelessness, substance abuse) that have accompanied large-scale oil sands development in Wood Buffalo. It is important to acknowledge these tensions for First Nation and Métis youth when thinking about their pursuit of education and work pathways. (Taylor et al. 2009:v).

Another form of engagement between Aboriginal communities and industry involves consultation efforts, the outcome of which includes traditional land use studies. But despite evidence of increasing employment and land use research part-nerships between Aboriginal groups and oil industry, some critics see Aboriginal employment and land use research collaborations as tactical interventions that help industry avoid public hearings and costly delays caused by them (Taylor et al. 2009:4, Note 8). Critical voices are wary of the robust claims made, and of ways in which the industry is conducting the partnerships that are in place. Some see part-nerships as free market incursions that erode nation status (Altamirano-Jimenez 2004). For others, traditional land use research appears a mixed outcome that if not carefully controlled can be misused by companies for instrumental problem solv-ing, such as establishment of setbacks, that allows operations to continue unabated, deeply miscomprehending or just plain ignoring Aboriginal knowledge, livelihood needs, and spiritual values of nature shared by the Elders. At its worst, corporate attempts at the indoctrination of Aboriginal people simply perpetuate stereotypes that serve to marginalize. According to Friedel, in her assessment of oil sands corporate advertising and newsletters, "the manner in which oil and gas transnationals' displaying of a racialized Native subject in the context of 'partnership' serves as a greenwashing strategy" (2008:238).

A Long Way from Home

The tar sands boom has transpired into a memorandum of agreement with both levels of government that allows companies to recruit foreign workers (AFL 2009a, b:14). Currently, about 40% of TFWs are categorized as skilled, and the "Low Skilled Worker" category makes up the other 60%, including construction laborers and service workers for retail, hospitality, cleaning and janitorial, transportation, domestic services, and child care. Most are from the Philippines, Mexico, and India. In February 2009 there were 57,843 TFWs in Alberta, a 55% increase over 2007, far outnumbering the 25,000 immigrants who arrived in Alberta in 2009 to become permanent residents. Just how many of these TFWs are employed in Ft. McMurray is unclear, but certainly a hefty proportion. Others work around the province at low-paying jobs vacated by Albertans who are work-ing the boom up north.

Western lands of opportunity such as that in the tar sands have the reputation of being available to all adventurous comers, but the most adventurous of all – those crossing national borders, and often oceans, to participate – face foreboding ceilings to their efforts at betterment. The most immediate ceiling is imposed by immigra-tion officials: TFWs arrive with a temporary work permit in hand, and limited hope for permanent residency or immigration. De Guerre (2009: 12) noted stark differ-ences across occupational sectors: "[s]killed professions [geologists, engineers, financial experts] are offered full recruitment and relocation packages that allow them to move their families and access to numerous benefits. [Un]-skilled trades are only granted the TFW permits for themselves, live in camp, and are not given the benefits 'packages.'" Most who apply to the Alberta TFW program suffer from the myth of big money and plentiful jobs, and come with hopes of sending large remittances back home, but few anticipate the expensive housing, unpredictable employment swings, and high cost of living. With the recent downturn, some are even reporting the need to ask their families to send money the other way. Other challenges include the cold, the darkness, the long winter, cultural isolation, language difficulties, small town life, and racism. The services available to meet the needs of an increasingly international population are overwhelmed. The AFL is strident in its condemnation of the TFW program, which it describes simply as racist and exploited by Government to please employers and facilitate a cheap, flexible workforce with fewer rights than ordinary Canadians. As the 2008 downturn hit, the AFL's analysts observed continued high rate of demand for TFWs, while permanent residents in high paying jobs were being laid off (AFL 2008:6). And the fact that TFWs tend to fill nonunionized jobs exacerbates the unequal relationship between corporations and labor for all workers. The questions raised by opposition MLA Backs to his colleagues in the Legislative Assembly were met with silence:

> For many years there was a special discriminatory tax placed on Chinese immigrants wanting to work in Canada. This tax was called the head tax. It was designed to ensure that Chinese workers on projects here would not bring their families to Canada and become Canadians. The special temporary foreign worker program for the oil sands negotiated and signed by the Alberta government has the same effect. My question is to the Minister of Human Resources and Employment. Why will Chinese temporary foreign workers contracted to work in the oil sands for, potentially, years, to live in work camps here for years not be allowed to have their wives and children immigrate to Alberta? (24 April 2006).

In sum, big paychecks come with big costs, not just in the cost of living, but also risks to health, risks to family relationships, the cost of silence about working conditions, the hazards of commuting on the highway of death, and for some, the risks to culture and way of life. TFWs face these same risks, but for them the rewards are even lower. There is slim chance for citizenship, constant fear of being shipped back home, higher risks of injury and constraints on service access due to language barriers, in a foreign and at times hostile cultural environment.

In many natural resource-dependent communities that have been the subjects of scrutiny by rural sociologists, corporations came to enjoy the quiescence of a local and relatively permanent work force that more often have not sided with the company in political debates, quieted by the power of proximity to employer. Today, the labor supplying such sites is nearly as global and as fluid as the commodity chains defining the raw materials themselves, as workers commute back and forth from near and far. The outcomes for community are rather clear, with the loss of human capital from communities of origin, and the lack of investment in the communities of employment. But the outcomes for staples politics are less certain. Temporary foreign workers are clearly in a subordinate position, and unions are being challenged on a regular basis in our neoliberal economies. On the other hand, mobile laborers, particularly those with specialized skills, are less dependent upon a single job-providing corporation. These individuals are also exposed to discourse, and social networks, outside of the extractive town, and not all of the information received is likely to be positive. To the contrary, saturated by media stories of the tar sands, the friends and families of workers back home in Edmonton, Calgary, Toronto, Newfoundland, or the Philippines may openly challenge their source of income and employment.

Our Fair Share

We know that wells that aren't drilled don't generate royalties ... and don't benefit Albertans. (Stelmach. Canadian Association of Drilling Engineers and Canadian Association of Drilling Contractors. Calgary, AB, 3 June 2008).

In the mid 1990s, 1% royalty provisions were developed, justified by high capital and operating costs, and low oil prices. Technically, companies would only pay this rate until all capital costs were recovered, and then pay 25%, but few companies have made it into this bracket as of this writing. The resulting "generic royalty regime" for both mining and in situ tar sands projects was the outcome of collaboration between the Alberta Chamber of Resources, the National Oil Sands Task Force and the oil sands corporations. That rate remained unchanged until criticism became heated in the mid 2000s, boosted by the well-publicized dialogue between Klein and Lougheed. By that time, it was obvious that the industry was well established and that oil prices were skyrocketing. In December 2006, newly elected leader Ed Stelmach followed through on one of his biggest campaign promises, and appointed the Alberta Royalty Review Panel (RRP), which held public hearings across the province (Plourde 2009), close on the heels of the Oil Sands Consultations. Those disproportionalities in the distribution of wealth from which proponents work so hard to divert our attention were not lost on many citizens:

You know, Imperial Oil, Exxon, Ford Motor, and all these companies on February the 2nd, 2007, revealed that for one year, one year, they made a profit of $39.5 billion for their shareholders. How much does Alberta have for its shareholders? And that's over a period of 33 years. [Referring to the Heritage Fund, with roughly $12 billion at the time]. Now, we cannot afford to give away what is ours. We must act like owners (Oil Sands Consultations, Edmonton, 3 April 2007).

Now, why am I speaking to you today and why do I think we need a new royalty structure? The system presently is extremely unfair. There is no trickle down of wealth to the ordinary population. Billions of dollars are produced, and yet since all these gallons or barrels of oil have been produced, children in schools have been in crowded classrooms. The vulnerable in our society, the disabled, seniors, the mentally ill, the homeless and the poor, all these people are Albertans. They should be sharing in the wealth, but for some reason the government doesn't have money to take care of the basic needs of the population, and I know it's not up to you to do this, but I think the government needs to hear this from the ordinary citizen. (Oil Sands Consultations, Edmonton, 3 April 2007).

After its deliberations, the Alberta Royalty Review panel members concluded in their report – *Our Fair Share* – that the Klein deal on royalties favored producers with capital depreciation and income tax reductions by both Alberta and the Federal government: "Industry got the breaks that it needed and asked for. Oil sands have since emerged as the dominant factor in Canada's energy future. The breaks given

ten years ago are still in place. Just as rebalancing was needed a decade ago …
rebalancing is needed again now." (Plourde 2009:136–7). The panel proposed a new
set of formulas designed to increase public royalties, impose a uniform regime on
both the mining and in situ producers, and to address public concerns for fairness.
The RRP Report was accepted and a number of recommendations adopted by government, but the Cabinet also disagreed in some key areas and issued their own
findings in a report called – *The New Royalty Framework* (Alberta Royalty Review
Panel 2007). University of Alberta economist Andre Plourde, one of the original
royalty review panel members responded: "the changes proposed by the (original)
Panel would have yielded yields … more favorable to the owners [the public],"
whereas the so-called rebalancing by Government actually reestablished, especially
as oil prices increase, the royalty distributions found in the old 1997 generic regime
under Klein, which were more favorable to industry (Plourde 2009:137). On a
personal note, Plourde had this to say:

> It's been a downer afterwards not so much because they said, well we're not going to do
> exactly what you said. But more the policy attitude. The fact that these are matters to be
> solved and resolved or addressed behind closed doors with no kind of public discussion,
> debate, this is a minister and a few other people have a chitchat with a few others and solve
> the problems. It says the wrong things about policy processes in democratic institutions, but
> that's a personal feeling. That, to me, is the depressing bit. The depressing bit is not so much
> that they didn't do what we said…although I still think what we said was a pretty reason-
> able thing to do and after much debate, discussion and turning it upside down, I still feel it
> was the right way to go, but the big depressing thing is kind of … everything was open and
> from one day to the next, everything was closed and back to business behind closed doors.
> I think that's the wrong way to do policy (Mouat 2009).

Premier Stelmach's defense? Corporations have our best interests in mind and
we should thank them.

> In uncertain economic times, it's our energy industry that is underwriting the prosperity of
> our country … attracting much-needed investment, and creating jobs from coast to coast.
> More and more, it is the energy industry and the Alberta economy that are supporting social
> programs—like health care—that Canadians value so highly. (Stelmach. Address to the 6th
> Annual TD Securities Oil Sands Forum, Calgary, AB, 9 July 2008).

His claims that "more drilling makes more royalties" avoids public questions
about royalty rates. Instead, the premier's rhetoric pounds home the simple message that the energy industry needs constantly to expand to benefit Albertans. For
the bankers, he raises the stakes by claiming Alberta's energy industry is under-
writing Canada's most sacred public service – heath care. Critical analysts label
such rhetorical techniques as "bridging effects," that is, ways of speaking that ties
tar sands oil (which has low social standing) to Canadian healthcare (a high pres-
tige discourse). The association does not always work. Journalist Sheila Pratt discovered that many Albertans are not against the energy industry per se, rather they
prefer more measured development – what she called a "reasonable prosperity"
(Pratt 2010). It is this sense of reasonable prosperity that we find in the public's
call for a slowdown in order to put environmental protections and social supports
in place.

Not surprisingly, while Plourde was lamenting the lack of democracy, the response from the oil and gas industry to the original report was "overwhelmingly negative" (Stikeman 2007), but they did not like Stelmach's amended report either. Syndicated financial analyst Don Coxe announced a scathing sanction from the financial sector, likening the modest redistribution to communism:

> We considered the (original panel review) report such a poorly-written, poorly-reasoned, mean-spirited betrayal of the traditions of a great province that we assumed it would be treated as an embarrassment. It failed to achieve even mediocrity, so it could be safely ignored ... Premier Stelmach stunned us by endorsing both the tone of the Panel's collectivist rant and most of its recommendations. Most importantly, he broke a promise he had made publicly not to accept its recommendation to break promises made to Suncor and Syncrude, the pioneers of the oil sands development ... With deep regret we are forced to remove Alberta from the shrinking list of politically-secure regions of the world for the oil industry, taking its rating from AAA to A (Coxe 2007).

Coxe's political insecurity rhetoric and ratings downgrade reveal the extent of cohesion between the energy and finance sectors within the capitalist elite: when neoliberalism is threatened, capital defends, even when the threat comes from one of its close partner states.

By way of defense, Alberta's politicians emphasized the complexity of formulas, characterized the public as incapable of understanding this complexity, and urged public trust in government wisdom. This member of public saw through the "ignorant public" character bashing:

> I mean the problem is not the people in this province don't understand our royalty regime, or the challenges that are faced by industry. I think they understand it better than a lot of people give them credit for. It's that they don't like it; that they're dissatisfied with it. So explaining it to them is just not going to do the job. (Oil Sands Consultations, Calgary, 18 April 2007).

The leader of the Liberal opposition party, Kevin Taft took advantage of the division between the conservatives and voters, alleging "preferential treatment of royalties required of the older mining companies Suncor and Syncrude," whose "royalties per barrel of bitumen decreased to just 48 cents a barrel – less than you'd receive by returning two 2-L pop bottles to the bottle depot" (Alberta Liberal Caucus 2009. Press Release, 12 May).

Taft's two pop bottle metaphor set in relief how Alberta royalty schemes were designed in an era of low price oil and high cost tar sands development, in which new in situ and SAGD extraction technologies were not anticipated, nor was the doubling and even tripling of the market price of nonconventional oil (Engemann and Owyang 2010). The Alberta Heritage Fund, established by Lougheed to set aside a portion of royalties, became the center of attention, particularly when placed in an international light. Alberta's fund stood at a miserly $14.4 billion after more than 34 years, while Norway's fund, for example, was greater than $374 billion after only 15 years. Even Alaska, a state that could hardly be described as a social democracy, also secured a royalty fund that put Alberta's to shame (Plourde 2009; Taylor 2006; Harder 2009).

> Other countries like Norway, Nigeria, and the state of Alaska get much more money for their oil, and with the exception of the despots in Nigeria and maybe Alberta, all are thriving

economies with great infrastructure and low unemployment as well as very healthy Heritage Funds, which puts ours to shame. This government needs a plan for the future that benefits all Albertans, not just those in the higher echelons of the oil industry. (Oil Sands Consultations, Calgary, 24 April 2007).

If development of the tar sands is allowed to continue at the rapid and seemingly unchecked rate that it has been, then this dream of a better world is not worth dreaming ... The fast pace and profit-motivated selfish way in which the oil industry pursues the American dream may have all sorts of benefits, but a future community full of worthwhile individuals capable of positive leadership is definitely not one of them. (Oil Sands Consultations, Edmonton, 3 April 2007).

It is interesting that in boom times, low royalties are justified by proponents on the basis of the fact that Alberta is making lots of money off the backs of industry (although the proportion of royalties paid vs. corporate profits is never mentioned), while in bust times or low oil prices, corporations come crying to the province that they cannot possibly pay high royalties or they will go under.

Where Alberta Stands

Alberta's take in the oil sands is lower than elsewhere in North America and is far lower than other regions of the world. These figures refer to various jurisdictions' take of energy dollars, including royalties, taxes, and other levies.

Alberta oil sands: 50%
UK: 53%
Newfoundland: 55%
California: 56%
Texas: 57%
Louisiana: 58%
1995 oil sands task force target: 60%
Ecuador: 72%
Nigeria: 85%
Venezuela heavy oil: 86%
Angola: 86%
Russia: 87%
Libya: 95%

Information sources from Alberta Department of Energy, Cambridge Energy Research Assoc., Wood MacKenzie, K.E. Andrews & Co. As reported in Ebner D (2007). As Big Oil pumps out profit, Alberta's take is shrinking. Is it time to up the ante? *Globe and Mail*, Saturday, Aug 18.

A Closer Look at the Cost Benefit Equation

> The decisions on the pace of oil development and how it is done is up to Albertans. There is no rush. The oil isn't going to evaporate, nor is peace going to spontaneously break out in the Middle East. It's our oil, we can develop it at our pace. Oil companies want our oil. They're not going to take their ball and go home if we change a few rules. (Oil Sands Consultations, Bonnyville, 10 April 2007).

Who is taking the risks, and who gains the benefits of past and present tar sands expansions? The risks borne by capital have been cushioned by the public purse since the first years of commercialization. Taxpayers develop industry roads, infrastructure, even labor policy and practices on behalf of – and in many cases not even at the behest of – the industry. Royalty patterns show that producers have benefited enormously from royalty formulae and tax write downs, and the owners of the resource are offered a fraction of the economic payoff. The risks to workers, by contrast, are multiple. Some make money but pay several prices for their paychecks, and many others suffer the consequences of inflation. The shifting of many jobs offshore or south of the border, or into the hands of temporary guest workers, flies in the face of proponents' assertions that Alberta jobs are the priority.

Some of the negative consequences of the boom are discussed in more hushed tones. The political discourse is all but silent on workplace drug and alcohol abuse and addictions, or concerns about health, fractured family life, loneliness, and lack of freedom to speak out about work life or work risks. Even less discussed is the future: how much mess will be left? What will be the costs of future cleanups? What is the likelihood that the energy corporations will stick around and chip in after profits dwindle?

References

Alberta Government. (2010). 2010 Fact Sheet Alberta Government: Talk About Oil Sands. http://www.energy.alberta.ca/OilSands/791.asp. Accessed 24 Jan 2011.

Alberta Government. (2010) Oil Sands Consultations Final Report. http://www.energy.alberta.ca/OilSands/pdfs/FinalReport_2007_OS_MSC.pdf

Alberta Energy Resource Conservation Board. (2008). Alberta's Energy Reserves 2007 and Supply-Demand Outlook 2008-2017. www.ercb.ca/docs/products/STs/st98-2009.pdf. Accessed 24 Jan 2011.

Alberta Environmental Law Centre. (2007) Oil Sands Consultations: A Backgrounder. http://www.elc.ab.ca/pages/Publications/PreviousIssue.aspx?id=570

Alberta Federation of Labour. (2003). Running to Stand Still: How Alberta government policy has led to wage stagnation during a time of prosperity. http://www.afl.org/index.php/View-document/16-Running-to-Stand-Still.html. Accessed 24 Jan 2011.

Alberta Federation of Labour. (2007). Anatomy of a Boom. http://programming.afl.org/index.php?option=com_content&id=826:anatomy-of-a-boom-2007&view=article&Itemid=154. Accessed 24 Jan 2011.

Alberta Federation of Labour (2008). Temporary Foreign Workers: Alberta's Disposable Workforce.   http://www.afl.org/index.php/Winter-2008/temporary-foreign-workers-albertas-disposable-workforce.html.

Alberta Federation of Labour. (2009, April). Lost Down the Pipeline: In these difficult economic times, is the Alberta government doing enough to keep value-added oil-sands jobs in Canada? http://www.afl.org/index.php/Download-document/116-Lost-Down-the-Pipeline-April-2009.html. Accessed 24 Jan 2011.

Alberta Federation of Labour. (2009, April 1). Entrenching Exploitation: Second Report of AFL Temporary Foreign Worker Advocate. http://www.afl.org/index.php/Reports/entrenching-exploitation-second-rept-of-afl-temporary-foreign-worker-advocate.html. Accessed 24 Jan 2011.

Alberta Royalty Review Panel. (2007). Our Fair Share—Report of the Alberta Royalty review panel. http://www.albertaroyaltyreview.ca/panel/final_report.pdf. Accessed 24 Jan 2011.

Altamirano-Jimenez, Isabel. (2004). North American First Peoples: slipping up into market citizenship? *Citizenship Studies 8*(4), 349–365.

Angell, A. (2010, October). Exploring the identity and wellbeing of mobile resource workers in the Alberta Oilsands. Paper presented at the Unwrap the Research Conference, Fort McMurray.

Bechtel Corporation (1998). Building A Century. 1970 — 1979 A decade of megaprojects. http://www.bechtel.com/BAC-Chapter-5.html. Accessed 25 Jan 2011.

Blake, M. (2010, October). Keynote Speech. Unwrap the Research Conference. Fort McMurray. Author's notes.

Bone, R. (1998). Resource towns in the Mackenzie Basin. *Cahiers de géographie du Québec 42*(116), 249–259.

Boychuk, R. (1996). *River of Grit*. Edmonton: Duval House.

Boychuk, R. (2010). Misplaced Generosity: Extraordinary Profits in Alberta's Oil and Gas Industry. Parkland Institute, Edmonton. Available at: http://parklandinstitute.ca/downloads/reports/MisplacedGenerosity-Web.pdf. Accessed 20 Dec 2010.

Calgary Herald. (2008, November 4). Tories "handing" U.S. oilsands upgrading jobs. *Calgary Herald* .

Canadian Broadcasting Corporation (Producer), & Pearson, P. (Director). (1977) *The tar sands* [Motion picture]. Canada: CBC.

Canadian Energy Research Institute. (2009). Economic Impacts of the Petroleum Industry in Canada: Study 120. http://www.ceri.ca/#oilsandsupdate. Accessed 24 Jan 2011.

Clarke, T. (2008). *Tar Sands Showdown: Canada and the New Politics of Oil in an Age of Climate Change*. Toronto: J. Lorimer & Co.

Cross, P. & Bowlby, G. (2006, September). The Alberta economic juggernaut: the boom on the rose. Statistics Canada, Canadian Economic Observer.

Coxe, D. (2007, November 7). Basic Points Newsletter. http://www.victoradair.com/pdf/Basic Points20071109.pdf. Accessed 24 Jan 2011.

de Guerre, K. (2009). Temporary Foreign Workers in Alberta's Oil Sector. MA Thesis. University of Sussex. http://www.sussex.ac.uk/migration/documents/mwp54.pdf. Accessed 24 Jan 2011.

Dembecki, G. (2010). Oil Workers Don't Cry. http://thetyee.ca/News/2010/08/16/OilWorkersDontCry/. Accessed 24 Jan 2011.

Edemariam, A. (2007, October 30). Mud, sweat and tears. *The Guardian*. http://www.guardian.co.uk/environment/2007/oct/30/energy.oilandpetrol. Accessed 24 Jan 2011.

Edmonton Journal. (2006, June 15). Fort Mac right to cry "enough!"

Engemann, K. & Owyang, M. (2010). Unconventional oil production: Stuck in a Rock and a Hard Place. http://research.stlouisfed.org/publications/regional/10/07/oil.pdf. Accessed 24 Jan 2011.

Epp, R. (1984). The Lougheed government and the media: news management in the Alberta political environment. *Canadian Journal of Communication 10*(2), 37–65.

Filax, G & Specht, A. (2009). Producing Alberta-ness in Texas North. Paper presented at Southwest Texas Popular Culture and American Culture Association Annual Conference, Albuquerque, NM.

Finkel, A. (1989). *The Social Credit Phenomena in Alberta*. Toronto: University of Toronto.

Gilbert, R. (2010). South Korean-made components a challenge for Kearl oilsands project. Journal of Commerce, 27 Oct. http://www.joconl.com/article/id41207/roadbuilding. Accessed 24 Jan 2011.

Gillmor, D. (2005, April). Shifting sands. *The Walrus*. http://www.walrusmagazine.com/articles/2005.04-alberta-tar-sands/. Accessed 24 Jan 2011.

Gray, J. (2008). The second coming of Peter Lougheed. *Globe and Mail*, 29 August.

Harder, A.C. (2009). Fuel and Fire Development versus Economic and Environmental Balance: What the Alberta Oil Sands Can Learn From the Norway Governance Model. MAIS Project. Athabasca University. http://library.athabascau.ca/drr/viewdtrdesc.php?cpk=0&id=39312. Accessed 24 Jan 2011.

Krahn, H. & Harrison, T. (1992). "Self-referenced" relative deprivation and economic beliefs: the effects of the recession in Alberta. *The Canadian Review of Sociology and Anthropology 29*(2), 191–209.

Kunzig, R. (2009, March). The Canadian oil boom. *National Geographic*. http://ngm.nationalgeographic.com/2009/03/canadian-oil-sands/kunzig-text. Accessed 24 Jan 2011.

Laxer, G. & Harrison, T. (Eds.). (1995). *The Trojan Horse: Alberta and the Future of Canada.* Montreal: Black Rose.

Lisac, M. (2010) Annual Review of Public Sector Salaries. Alberta Washington Office. Insight Into Government. *25*(7).

Lizée, E. (2010). Rhetoric and Reality: Albertans and Their Oil Industry Under Peter Lougheed. MA Thesis. Department of History and Classics, University of Alberta. http://repository.library.ualberta.ca/dspace/bitstream/10048/1147/1/Erik%2BLizee%2BSpring%2B2010.pdf. Accessed 24 Jan 2011.

Maich, S. (2005, June 13). Alberta is about to get wildly rich and powerful. What does that mean for Canada? *Macleans*. http://www.macleans.ca/business/companies/article.jsp?content=20050613_107308_107308. Accessed 24 Jan 2011.

Makin, K. (2007). Clash over oil sands inevitable: Lougheed. *Globe and Mail*, 14 August.

Mansell, R. & Schlenker, R. (1995). The Provincial Distribution of Federal Fiscal Balances. Canadian Association of Business Economics. http://www.cabe.ca/jmv1/index.php?option=com_docman&task=doc_download&gid=124&Itemid=38. Accessed 24 Jan 2011.

Marsden, W. (2007). *Stupid to the Last drop: How Alberta is brining Environmental Armageddon to Canada (and doesn't seem to care).* Toronto: Random House.

Marsh, J. (2005). Alberta's Quiet Revolution. In Payne, M., Wetherell, D., & Cavanaugh, C. (Eds.). *Alberta Formed Alberta Transformed.* Edmonton: University of Alberta. pp. 642–674.

Mayes, M. (2009). Ed Stelmach is soiled by magazine's Oilsands Coverage. The Edmonton Journal. 25 February.

Mouat, J. (2005). Interview With Larry Pratt and John Richards, Authors of Prairie Capitalism, 25 Years Later. *Aurora Online*, available at: http://aurora.icaap.org/index.php/aurora/article/view/8 Accessed 24 Jan 2011.

Mouat, J. (2009) Interview with Andre Plourde. *Aurora Online* In archives.

Nikiforuk, A., Pratt, S., & Wanagas, D. (1987). *Running on Empty: Alberta After the Boom.* Edmoton: NeWest.

O'Shaughnessy, S., & Krogman, N. (2010), Women's diverse experiences of rapid resource development in Fort McMurray. Paper presented at Unwrap the Research Conference, October, Fort McMurray.

Owram, D. (2005). Oil's Magic Wand. In Payne, M., Wetherell, D. & Cavanaugh, C. (Eds.). *Alberta Formed.* Edmonton: University of Alberta.

Paehlke, Robert C. 2008. Some like it Cold: The Politics of Climate Change in Canada. Toronto: Between the Lines.

Palmer, H. with Palmer, T. (1990). *Alberta: A New History.* Edmonton: Hurtig.

Pasqualetti, M. (2009). The Alberta Oil Sands from Both Sides of the Border. *The Geographical Review* (April), 248–2671.

Plourde, A. (2009). Oil Sands Royalties and Taxes in Alberta: An Assessment of Key Developments since the mid-1990s. *The Energy Journal 30*(1), 112.

Pratt, L. (1976). *The Tar Sands: Syncrude and the Politics of Oil.* Edmonton: Mel Hurtig Publishers.

Pratt, L. (1977). The state and province building: Alberta's development strategy. In Panitch, L. (Ed.). *The Canadian State: Political Economy and Political Power.* Toronto: University of Toronto.

Pratt, L. & Urquhart, I. (1994). *The Last Great Forest: Japanese Multinationals and Alberta's Northern Forests*. Edmonton: NeWest.

Pratt, S. (2010). A line in the oil sands. *Vancouver Sun*, Section C, 1, 6 Nov.

Prescott, M. (2009). *Fort Mac*. Saint-Boniface, Manitoba: Les éditions du blé.

Preville, P. (2006). Down and dirty in Fort McMurray. *Chatelaine 79*(Oct), 10.

Reguly, E. (2006, May 26). Boom gone berserk. *Globe and Mail*.

Richards, J. & Pratt, L. (1979). *Prairie Capitalism: Power and Influence in the New West*. Toronto: McClelland and Stewart.

Sallot, J. (1982, May 11). CBC to pay Lougheed $82,500 to settle lawsuit over oil sands drama. *Globe and Mail*.

Shiell, Leslie, and Suzanne Loney. 2007. Global Warming Damages and Canada's Oil Sands. Canadian Public Policy 33 (4):419–440.

Sinclair, P. (2011). *Energy in Canada*. Oxford: Oxford University.

Slowey, G. (2008). *Navigating Neoliberalism: Self Determination and the Mikisew Cree First Nation*. Vancouver: UBC.

Spragins, F. Personal Papers, Glenbow Museum. M 5873 M 5888 NA 4173. Typed biography co-written by Spragins, F. with Barr, J. 1978–1980.

Soron, D. (2004). The cultural politics of Kyoto: lessons from the Canadian semi-periphery. *Capitalism Nature Socialism 15*(1), 43–66.

Statistics Canada. 2006 Census. Aboriginal Labour Characteristics.

Stewart, G. (2007, 23 August). Rudderless in the oilpatch: Alberta's pending political storm. *Fast Forward Weekly*. http://www.ffwdweekly.com/article/news-views/viewpoint/rudderless-oil-patch/. Accessed 24 Jan 2011.

Stikeman, E. (2007). Canada's Business Law Firm. http://www.stikeman.com/cps/rde/xchg/se-en/hs.xsl/10381.htm. Accessed 24 Jan 2011.

Storey, K. (2010, October). The Big Commute. Living in Newfoundland and Working in the Oil Sands. Paper presented at Unwrap the Research Conference, Fort McMurray. http://www.uofaweb.ualberta.ca/crsc/pdfs/Unwrap_-_The_Big_Commute.pdf. Accessed 24 Jan 2011.

Strategy West Inc. (2009). Existing and Proposed Canadian Commercial Oil Sands Projects, February. http://www.strategywest.com/oilSands.html. Accessed 24 Jan 2011.

Sweeny, A. (2010). *Black Bonanza: Canada's Oil Sands and the Race to Secure North America's Energy Future*. Mississauga, ON: Wiley.

Taft, K. (1997). *Shredding the Public Interest: Ralph Klein and Twenty-Five Years of One-Party Government*. Edmonton: University of Alberta.

Taylor, A. (2006, March 25). Klein Fails to provide nest egg. *Edmonton Journal*.

Taylor, A., Freidel, T., & Edge, L. (2009). Pathways for First Nation and Métis Youth in the Oil Sands. Canadian Policy Research Networks Report.

Vicente, M.E. (2005). The National Energy Program. Canada's Digital Collections. Heritage Community Foundation. http://www.abheritage.ca/abpolitics/events/issues_nep.html. Accessed 24 Jan 2011.

Weber, B. (2010, October 11). Alberta's oilsands city wants workers to live in, not just cash in. The Canadian Press, as found in *680 News*. http://www.680news.com/news/national/article/113580--alberta-s-oilsands-city-wants-workers-to-live-in-not-just-cash-in. Accessed 24 Jan 2011.

Weir, B. (2008, August 14). Canada has green concerns over oil goldmine. ABC News. http://abcnews.go.com/Business/story?id=5464319&page=1. Accessed 24 Jan 2011.

Chapter 5
Ecological Disruption

> *Proving to our customers that Alberta's energy products are secure and clean will be the challenge of this decade. Government and industry will have to work together to provide factual evidence, so that public opinion is informed about all the measures we're taking on the environment. We're doing a better job of environmental protection here than people think. We need to tell Alberta's environmental story* (Stelmach, Address to the Canadian Association of Drilling Engineers and Canadian Association of Drilling Contractors, Calgary, June 3, 2008).

Canada's boreal forest ecozone is a major part of the global boreal region that encircles the Earth's northern hemisphere, serving as a significant storehouse for the world's freshwater supplies, and carbon, contained in its trees, soil, and peat (Schneider and Dyer 2006:1). The boreal is also home to a rich array of wildlife, including migratory birds, waterfowl, bears, wolves, moose, and caribou. Alberta's share of this boreal forest covers 381,000 km² (Alberta Environment 2009:5). Through this runs the Athabasca River, which, at 1,538 km "is Alberta's longest river and one of the few free-flowing (undammed) rivers left in North America" (Woynillowicz and Severson-Baker 2006:ii), and the Peace-Athabasca Rivers watershed has been identified as the World's third largest watershed (after the Amazon and Mississippi Rivers).

Flows of resources, capital, people, and discourses through this watershed leave a multitude of ecological wakes. Many of these wakes are latent, not revealing themselves until years after the flows have run their course. Others present themselves immediately but are unrecognized as such. Still others are evident but their meaning, and by extension our responses to them, are contested. While all categories are equally relevant to ecosystem wellbeing, we turn our attention to the last. These wakes have entered the realm of politics today, and the means by which they are processed in our political system reveal society's relationship to the natural world in vivid colors. International discursive attention to the ecological footprint posed by the tar sands has only emerged within the past 10 years, coinciding with the dramatic increase in production experienced during this time frame. The key vehicle for this

D.J. Davidson and M. Gismondi, *Challenging Legitimacy at the Precipice of Energy Calamity*, DOI 10.1007/978-1-4614-0287-9_5, © Springer Science+Business Media, LLC 2011

Fig. 5.1 A braided portion of the upper Athabasca River. Reproduced with the permission of Natural Resources Canada 2011, courtesy of the Geological Survey of Canada (photo 2001-348 by Denis St-Onge). http://atlas.nrcan.gc.ca/site/english/maps/environment/land/physio_athabasca_river.jpg. Accessed January 15, 2011

Fig. 5.2 The lower Athabasca River. Fort McKay. With permission of Louis Helbig. B2402294. See www.beautifuldestruction.com and http://www.louishelbig.com/

Fig. 5.3 Boreal Forest and Mist. 50 km west of Fort McMurray. With permission Louis Helbig. See www.beautifuldestruction.com and http://www.louishelbig.com/

discourse has been the visual image, including photographic and satellite imagery conveyed readily to global viewers via the electronic universe (Figs. 5.1–5.3).

Three ecological wakes in particular define the environmental politics of tar sands development: landbase disturbance, water use, and climate change. In the case of the first, tar sands development may well be the largest-scale landbase disturbance for a single industrial operation the world has known. As well, the enormous demands on regional water systems posed by tar sands operations have significant implications for regional water supply and quality, and also for a globally significant riparian ecosystem. Finally, the Athabasca tar sands contribute a relatively small amount of greenhouse gases from a global perspective today. However, both the anticipated exponential future growth rates in bitumen production and its associated emissions, *and* the symbolic relevance of increasing reliance on fossil fuel resources with *higher* greenhouse gas intensity than conventional sources demands close scrutiny. We begin this chapter with a brief synopsis of information available regarding the ecological impacts of tar sands development, followed by critical inquiry into the discourses that have evolved around these issues.

Footprints in the Tar

Getting to the Resource

The ecological impacts in evidence today begin with the complete removal of a layer of flesh from the earth's surface several meters deep and several thousand square kilometers wide. As of March 2009, 602 km^2 of terrestrial ecosystem had

Fig. 5.4 Overburden removal for mining operations. Louis Helbig. Beautiful Destruction overburden removal B2410245

been removed by mining (Alberta Energy 2009). An additional 4,000 km^2 has already been leased for potential future mining development over the next 10 years (Alberta Energy 2009b). The total area recognized as containing bitumen deposits is 142,200 km^2 (Alberta Environment 2009:5), so full development of the resource would encompass roughly 20% of the entire provincial landbase of Alberta. Land disturbed by in situ operations is a bit harder to quantify – an individual well pad consumes between 4 and 4.5 hectares of land (each well pad will have several wells), but servicing these well sites requires an extensive road and pipeline network, posing a significant degree of fragmentation on the land base, and thus rendering it effectively unavailable for the support of ecological systems. There were 87 in situ wells in operation as of August 2009 (Alberta Energy 2009); this number will increase significantly however; over 80% of the resource can only be accessed with this process (ibid). Approximately 80,000 km^2 have been leased for future in situ development (Alberta Energy 2009b) (Fig. 5.4).

Pipelines, which function similar to roads in terms of fragmentation, are also required to move the fuel from sites of extraction to sites of upgrading and refining. Current estimates of the linear extent of this pipeline network in Alberta alone is 412,555 km (Alberta Energy 2010a); inclusion of pipelines used to transport raw and upgraded bitumen for further processing and final distribution south of the border would increase this figure substantially.

Approval of new operations is premised on assurances that the land will be reclaimed upon completion of the industrial operation. Once all the mineable oil sands in an area have been removed, the mine pit is slated to be converted into an "end pit" lake, which, according to operators, will become a viable aquatic ecosystem. Land subject to other uses is slated for land reclamation, including tailings ponds currently covering 50 km², which is where water currently too toxic to be released back into the river is stored.

What Goes In

The next layer of impact entails the resources that go into the extraction and processing operations. Energy inputs will be discussed in the next chapter; here we focus on the other key input: water. Water used includes fresh surface and underground sources, and saline water from underground sources. Water that is presumed to be sufficiently clean after industrial use is usually recycled, with some companies claiming to recycle up to 90% of water consumed several times over. According to the industry's own association, however, the Canadian Association of Petroleum Producers, oil sands operations use approximately 176 million cubic meters of water per year, 160 million of which are drawn from the Athabasca River (CAPP 2010), so the 90% recycling figure clearly does not represent water consumption practices for many operators.

Industry and state representatives are quick to point out that current water consumption from the river amounts to less than 1% of average total river flows, or five per cent of flows during the low-flow winter months (CAPP 2010; Alberta Environment 2009). Flow rates vary dramatically on a seasonal and annual basis. However, considering that industrial operations have license to continue to remove water from the river regardless of flow rates, the ability to maintain ecological function is of particular concern during low flow periods. Climate change forecasts suggest declining water availability and increased intensity and duration of droughts raise the spectre of severe water shortages in the coming years. This limited and uncertain water supply is expected to support not just current tar sands operations, but also planned expansion up to a tripling of current development over the coming decade. The 80% or so of the resource that is too deep to be mined and thus accessed via varying well drilling techniques (and will be to a much greater extent in the future) use largely groundwater. Little is known about groundwater sources in northeastern Alberta, either how much is there, or whether current techniques pose a contamination risk to fresh groundwater sources.

What Comes Out

The stuff that comes out of the industrial processes ends up not just in our gas tanks but also in the water and air, some of which is permitted (allowable discharges), some of which are accidental (pipeline breaches, spills, facility malfunction events), and many of which are not "permitted" but nonetheless quite routine, including leakage and evaporation from the tailings ponds. Alberta's current system for managing water begins with an allocation system based on the first in time, first in right principle, with numerous water allocations granted in perpetuity. Beyond direct consumption, industrial operators must apply for a license for development that impacts a water body, under the Water Act. The government has attempted to address growing public concerns for water with development of the *Water for Life Strategy* earlier this decade, which sets the goal of 30% efficiency improvement in water use by 2015. The Province has invested directly in a groundwater monitoring system, but for the most part the provincial government relies on self-monitoring and reporting by the energy industry.

According to a growing number of accounts, the current regulatory structure has limited effectiveness. Among the nastier of outputs that end up in the water as a result of tar sands activities are polycyclic aromatic hydrocarbons (PAHs), mercury, and napthenic acids. PAHs include a number of compounds, many of which are suspected or known carcinogens, and increasing concentrations have been evidenced in the Athabasca River downstream from extraction operations (Timoney and Lee 2009). The health hazards of mercury – released when vegetation is flooded – are well-known, particularly their tendency to bioaccumulate through the food chain. Recent studies have confirmed increased mercury levels in the local Walleye population (Timoney and Lee 2009). Naphthenic acids occur naturally as a constituent of bitumen, but they become concentrated in mining tailings. Significant attention has been drawn to naphthenic acids of late: they tend to persist, and are quite toxic at concentrations found in tailings ponds. Concentrations of naphthenic acids in rivers within the Athabasca oil sands region are generally below 1 mg/L, but may be as high as 110 mg/L in tailings waters (Grant et al. 2008). The expectation (or wildly unfounded hope, depending on one's perspective) is that these and other compounds that result from the industrial process will settle to the bottom of tailings ponds, and be disposed of as a solid, with the water returned to the river. Actual rates of settlement have not come close to meeting expectations however; the latest studies suggest it could take as long as 100 years for the necessary settlement to occur (Grant et al. 2008:38), a long period of time for large, open, and highly toxic water bodies to be managed – water bodies that are already strongly suspected of leaking rather large contents into the Athabasca River. Such suspicions are not difficult to fathom, even for the untrained – a number of tailings ponds are separated from the Athabasca River solely by a few meters of dyke constructed from sand and clay. Some observers have also noted a suspicious lack of correlation between the reported volume of tailing effluent entering some ponds, and the lack of associated increase in pond volume. According to Toronto-based Environmental Defense, tailings ponds leak 11 million liters a day into the river, based on industry's own estimates.

A study conducted for Fort Chipewyan Community Health Authority indicated increases of mercury levels in Lake Athabasca of 98% since the inception of tar sands development (Timoney 2007). Arsenic levels in the lake water had also risen, by 466%, lead by 114% and PAHs by 72% (ibid). Both tar sands operators and the

Fig. 5.5 Bird deterrent technology in use on tar sands tailings ponds ranges from glorified scarecrows to robo-hawks equipped with radar detection. Photos courtesy of Colleen Cassady St. Clair

Government of Alberta, not surprisingly, disputed these findings. Proponents are quick to highlight the multiple other possible factors that may contribute to such consequences, particularly the "natural seepage" that occurs from the tar deposits. On the other hand, the toxicity of the ponds themselves is far more difficult to dispute. Migratory birds landing on the surface of the tailings ponds, which appear as attractive reprieves for these world travelers, die immediately. Five hundred or so (more recent estimates are as high as 1,600) died in 2008, when the failsafe bird deterrent system operated by Syncrude, well, failed (Fig. 5.5).

The gunk that gets pumped into the air is of concern as well. The air pollutants in question are typical of industrial operations, including sulfur and nitrogen oxides, ammonia, ozone, carbon dioxide, carbon monoxide, PAHs and fine particulate matter (Alberta Environment n.d.). Air quality standards in the province are not atypical for North America, but the level of compliance may be. Analysis of air quality tests collected by Suncor's own monitoring organization show that companies breached Alberta's air pollution targets more in 2009 than ever before, according to Environmental Defence Canada. Their recent report found companies violated the province's air quality standards more than 1,500 times in 2009, compared to less than 50 times in 2004 (Environmental Defence 2010:3). What goes into the air also has a tendency to end up in the water, as shown by Dr. David Schindler and his colleagues in a recent article (Kelly et al. 2009). These scientists found elevated levels of PACs – suspected to be toxic to fish embryos – deposited into the snowpack within the airsheds of upgraders, awaiting settlement into the Athabasca River during spring snowmelt.

As for the pollutant that has received tremendous fanfare lately, direct CO_2 emissions from the tar sands accounted for about 5% of Canada's overall emissions by 2008 (Alberta Energy 2009), the most recent year for which the data is available. Proponents are quick to point out that this represents a small proportion of global emissions, and rightly so. There are other facts and figures to consider, however. First, if political advocates on both sides of the border are successful, we could experience a fivefold increase in production rates in the coming decades. Secondly, since, as with all extractive operations, the easiest stuff is removed first, the energy inputs required for extraction and processing will increase over time. Both pose ever-escalating counter trends for technological innovators celebrating achievements in greenhouse gas *intensity* reductions. Finally, there are multiple hidden sources of emissions associated with tar sands development. Emissions are also associated with refining, and eventual consumption as gasoline, both of which occur primarily in the United States and tend not to be included in carbon footprint tallies for the tar sands. Operations in remote places also require the movement of materials and people up to the extraction sites in rather large quantities, and in the case of the latter, back out as well, since many workers commute on a regular basis to and from places like Newfoundland along newly established flight paths, or at least Edmonton and Calgary, usually in large trucks or SUVs, the work vehicles of choice for most. The very removal of the boreal forest, which when intact serves as a carbon sink, also has a negative impact on global carbon budgets, and it has been found that bacteria present in tailing ponds emit methane, a potent greenhouse gas. On the World Wildlife Foundation (WWF)'s climate change report card, released on July 1 2009, Canada ranked at the bottom of the list, and WWF argued that Alberta's tar sands are largely to blame (Figs. 5.6 and 5.7).

The impacts of pollution are global in scope, but are nonetheless felt most acutely at the local level, and no amount of scientific data can capture these costs as effectively as can local residents themselves:

> The ammonia leak last year, I was doing a school visit that day, and I came out, and there was some sick kids, there were some scared kids, there were adults that were scared. We had a situation at the day care where they filled the water table with snow this winter. When it melted, it smelled like sulfur and there was a black ring around the water table, so the kids can't play with the snow anymore, because it's not safe, yet the snow is still white. So there's lots of impacts that the environment has on our children that we don't even see (Oil Sands Consultations, Fort McMurray, March 27, 2007).

The local residents expressing "anectodal" concerns included the resident family physician in the First Nations community of Fort McKay, whose expressions of concern about what appeared to be elevated cancer rates were met with formal complaints from the Alberta College of Physicians and Surgeons of causing undue alarm. He has since been doubly vindicated, with complete clearance of charges, and confirmation of his suspicions in a study by the Alberta Cancer Board, released in February 2009.

Fig. 5.6 Sound Cannon N 56.54.51 W 111.23.04, Suncor Millennium Mine, Alberta, Canada. With permission. Layers of thick Bitumen surround a sound cannon and other equipment at the edge of a tar pond. The sound cannons are designed to deter birds from landing in this toxic water. www.louishelbig.com/

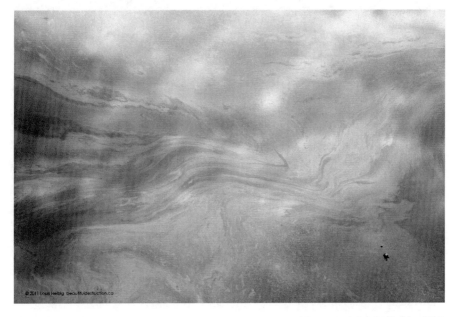

Fig. 5.7 Bitu-Man – A bird's eye view of a scarecrow on a tailings pond. Louis Helbig. With permission. See www.beautifuldestruction.com and http://www.louishelbig.com/

Land

The spatial scale and visual impact of the land disruption caused by tar sands development has most recently been referred to as a "gigaproject," "megaproject" having been deemed wholly inadequate to the scale of disruption. Other more qualitative imagery abounds, with Council of Canadian's Chair Maude Barlow referring to it as "Canada's Mordor," in reference to the fabled hell hole in Tolkein's *Lord of the Rings*, and analogies made to the devastation wrought in the film *Avatar*. The ready availability of imagery via the worldwide web has likely contributed to the internationalization of alarm, although many testimonials still suggest that "you have to see it to believe it." To the extent this is true, both proponents and opponents have gone to great lengths to characterize the scale of impact in a manner that suits their interests, through the use of numbers and metaphorical comparisons in discourse, and the careful selection of visual imagery that situates the development in a suitable light (Figs. 5.8 and 5.9).

The provincial government and corporations, on the other hand, appear primarily concerned about the allegedly cumbersome approval process for new operations. Representatives from both have in several iterations attempted to streamline this process by means of "integrated land use management." The Mineable Oil Sands Strategy, or MOSS, is one iteration that received a heightened level of critique when

Fig. 5.8 Aerial view of Syncrude Aurora tar sands mine in the Boreal forest north of Fort McMurray, northern Alberta. Image offered by Greenpeace to depict the "clearcutting" of the boreal forest for tar sands. Available at: http://www.greenpeace.org/canada/en/campaigns/tarsands/. Accessed January 17, 2011. © Jiri Rezac/Greenpeace; ID:GP026NU; 07/20/2009

Fig. 5.9 Heavy hauler. Image courtesy Syncrude Canada Ltd

introduced in late 2005. MOSS effectively develops land use zones, committing 2,600 km^2 to tar sands operations, nullifying the usual gamut of environmental regulations governing land use. According to the former Environment Minister, this prioritization of industrial activities will "ensure, number one, that the environment is protected and, number two, that necessary regulation is put in place to prevent unnecessary delays" (Boutilier, Legislative Assembly, November 23, 2005). Years earlier, in an attempt to allay the concerns of several stakeholder groups, in 2000 the Province established the Cumulative Effects Management Association (CEMA), including representation from industry, the Government of Alberta, Canada, the Regional Municipality of Wood Buffalo (where Fort McMurray is located), non-governmental organizations, and First Nations and Métis peoples. After several frustrating years of inaction, the majority of environmental and Aboriginal representatives pulled out, leaving the Association to the whims of state and industry interests. In essence, the primary land management vehicles in place are dominated by state and industry interests.

In addition to land use management, Alberta Environment oversees land reclamation, requiring all companies that disturb land to submit a reclamation plan prior to project approval. Provincial policy requires that disturbed land be reclaimed to a "productive" state. Lands that have been shown to be productive after reclamation are awarded a certificate. Operators are required to provide a financial retainer, held in trust by the government, as security for future reclamation costs. Interestingly, reflecting the extent of trust the government ascribes to multinational corporations, the companies themselves calculate what the deposit should be. In spring 2009, Alberta Environment held approximately $820 million in security from oil sands companies (Alberta Environment 2009:10).

Despite the stated policy of corporate liability for reclamation costs, the provincial government provides several forms of subsidization. The Government of Alberta recently invested $2 million towards reclamation research for oil sands projects, and they provide resources for the Cumulative Environmental Management Association. The most significant form of subsidy, however, is posed by the differential between the reclamation security and the actual costs of reclamation. The first (and currently the only) reclamation certificate was issued in March 2008, for 1 km^2 of land, Syncrude's Gateway Hill. Syncrude spent $30.5 million to reclaim this plot of land – $114,000 per hectare – which is located on upland (as opposed to wetland) and while in use, simply stored "overburden" and thus was not subject to mining or tailings (Hildebrand 2008:8). Mind you, the company undoubtedly spent an exaggerated amount on this particular reclamation project, the first to achieve a reclamation certificate from the Province, and their investment has paid off several times over in the form of public relations, not just for this company but for the industry as a whole. By contrast, the Security Fund held in trust by the Province has enough in it to provide roughly $10,000 per hectare for the landbase currently disturbed (ibid). Regardless of cost, there is a tremendous level of speculation about the extent to which we have the needed ecological ingenuity to perform successful land reclamation at all. Nikiforuk (2008:102), one among many skeptics, considers reclamation as "little more than putting lipstick on a corpse."

Voices of Praise

The spectacular scale of the tar sands operations was initially lauded by supporters, who presented this "mega" status as a source of pride, as discussed in Chap. 3. Proponents were seemingly caught by surprise when this very mega status became a mark not of pride but of prejudice. This challenge was met by expressions of assurance in the reclamation process, and the goodwill of the "corporate citizens" engaged in tar sands development, all the while attempting to reassert control over the meaning of "mega" as marvel rather than grotesque. This was captured most succinctly by the province's former Premier, Ralph Klein, who responded to questions posed by a radio talk show host in 2005:

> When you look at it you have to see it to believe it because the shovels, you know one shovel-ful is the size of this studio. So you can imagine the immensity of the amount of soil that is taken out.... So, it leaves a huge scar on the land, huge, because these, first of all, they mine very deep and they're open pit mines, but there are very strict reclamation measures. In order to see the kind of reclamation that takes place, one needs to travel to the north and see the work that Syncrude.... has done to restore an old mine site to its natural state and of course

they run buffalo on that land and there are deer. You would never know that it had been mined at one time (Klein, Interview with Mary O'Driscoll, OnPoint. March 29, 2005).

Over the years, members of the Progressive Conservative Association of Alberta (PC Party) have become seasoned defenders, although incidents such as the mass duck killing in 2008 certainly challenge their discursive skill. The exchange in the Legislature between Premier Stelmach and a member of the Opposition suggests a denial of culpability. Stelmach characterizes the killing as an isolated incident, provides assurances that justice will be served at the hands of a government which is still in the driver's seat, and evokes an unshaken reliance on our ability to control nature through the wonders of science:

Mason: Will the Premier admit that his government's failure to insist on dry tailings puts Alberta at an even greater risk of environmental catastrophe?

Stelmach: Mr. Speaker, we have the strictest legislation in place in the country … We'll continue to be very vigilant. For this particular incident that has caught so much attention, at least from the opposition, I said that we'll have a full inquiry…If there were any errors committed by whoever we issued the permit to, we will bring those people forward. We have very strict fines in place as well. This isn't a funny matter. This is serious, and we take it seriously. If there's any wrongdoing, all Albertans will be made very well aware of what happened at that site.

Mason: Well, it's great that the Premier is closing the barn door now that the horses are out, but I want to ask the Premier whether or not he will do the environmentally responsible thing and act to clean up the Athabasca tailings ponds with no further delay.

Stelmach: There is a lot of research and innovation being put in place in terms of the tailings ponds. There are a number of good ideas … I'm looking forward to more information coming out of further research (Legislative Assembly, May 7, 2008).

PC Party members are quick to point to the Syncrude buffalo paddock, a mere 1 km^2 in size but huge in its symbolic representation of our ingenuity, as well as corporate goodwill:

We have to have some justifiable pride in some of the companies that have… brought forward some great strides, I think, in ensuring that the land is reclaimed. If anybody goes near some of the areas in the oil sands and sees some of the little parks that are being developed, some of the buffalo paddocks, I think that they will agree that there are some great efforts being made (Backs, Alberta Legislature, May 16, 2006).

Corporations rely extensively on this support from the Legislature. Like the former Premier, industry reps are quick to acknowledge that "disturbance to the land is a reality with resource development,"[1] while also suggesting that a return to a functioning boreal ecosystem is a pipedream, and by extension, those espousing such standards are dreamers:

The idea of restoring disturbed areas to a so-called "natural state," which could mean different things to different stakeholders, would simply be difficult to define and enforce. Rather, Imperial suggests that the long-term reclamation goal should be to return these lands to an equivalent land capability acceptable to stakeholders (Lui, E.L. Imperial Oil, Written submission to Oil Sands Consultations, April 24, 2007).

---

[1]Chub, Derek, Suncor, Oil Sands Consultations, Fort Chipewyan, Oct 4, 2006.

Fig. 5.10 Image Courtesy Syncrude Canada Ltd. In Syncrude's own words: "Gateway Hill: The public is welcome to see land reclamation themselves on the 4.5-km interpretative trail through Gateway Hill. It's called Matcheetawin Trail, which is Cree for beginning place" (available at: http://www.syncrude.ca. Accessed July 25, 2010)

Rather than dwelling on what cannot be done, industry spokespersons emphasize what *can* be done, regardless of the fact that such accomplishments may have little to do with ecology. This discourse reflects a masterful ability to *create* accomplishments, as in the following speech excerpt, in which an industry rep heralds the fact that once mine expansion ceases, the rate of reclamation will exceed that of disturbance (mind you this would be true of any level of reclamation above zero) (Fig. 5.10):

> At our Base Mine site, the pace of annual land reclamation now exceeds the pace of distur-
> bance. This trend will continue as the mine reaches the end of its production life (Syncrude,
> Written submission to Oil Sands Consultations, n.d.).

The buffalo, not originally from the area, has been embraced as the symbol of Nature's vitality, with certification of such vitality expressed in rather ironic forms:

> More than 300 wood bison now graze on land reclaimed from our mining and tailings
> operations.... At the 2005 Wild Rose Classic bison show awards were won by each of the
> Syncrude bison entered including the top prize of Reserve Grand Champion (Syncrude,
> Written submission to Oil Sands Consultations, n.d.).

Industry nonetheless acknowledges that improvements can be made, and *will* be made if we just get the right people working together, namely industry and its supporters in government (Fig. 5.11):

> It is widely believed that the pace of reclamation for oil sands mines is too low and too slow.
> There are many reasons for this situation, but industry and government agree that we need

Fig. 5.11 Image courtesy Syncrude Canada Ltd

to work together to demonstrate better progress. We should also recognize that the regulatory processes are in place to deal with the issues (Doucet, Horizon Oil/CNRL, Oil Sands Consultations, Fort McMurray, March 28, 2007).

Voices of Discontent

Reclamation has not captured a significant amount of attention from international environmental organizations. Regional ENGOs, however, have discussed it at great length. Regional ENGO reps have picked up on this scientific uncertainty theme and used it as a basis for invoking the precautionary principle, while also calling into question the acceptability of the "equivalent land use" clause. As a representative of the Alberta Fish and Game Association put it, "The last thing Alberta needs is more artificial grasslands and bison ranches where native boreal forests once stood."[2] The Environmental Law Centre (ELC) offered a more extensive critique, one that represents well the discursive approach taken by a number of regional organizations:

---

[2]Boyd, Andy, Oral testimony at Oil Sands Consultations, April 4, 2007, Edmonton.

The ELC considers reclamation to a self-sustaining boreal forest ecosystem to be a laudable goal; however, as an action, this creates a requirement that is difficult to enforce effectively. The boreal forest ecosystem within which the Athabasca oil sands are wholly situated is a complex ecosystem composed of interconnected forests and wetlands. There is no certainty that this complex ecosystem, with its abundant biodiversity, can be recreated to a level that is self-sustaining. In the absence of a certain ability to recreate the complex boreal forest ecosystem, it is not sufficient for operators to simply reclaim disturbed lands by creating fenced livestock pastures or dry forested hills dotted with end-pit lakes (Chiasson et al. Environmental Law Centre. Written submission to Oil Sands Consultations, April 24, 2007).

Many storylines offered by provincial environmental organizations suggest a conservative discursive strategy designed to make incremental inroads into provincial policy. The ELC, for example, in the same written submission referred to earlier, explicitly qualifies their critique, stating that "The ELC does not necessarily consider that the suspension of oil sands activity is required until industry has 'caught up' with reclamation." But there are exceptions, particularly among smaller local organizations, the members of which appear less concerned about stepping on the toes of power. The following clips from citizen testimonials offer a less impersonal tone, while offering a more direct challenge to the legitimacy of the provincial government:

Just be honest with the public and state the facts. Reclamation is not possible for much of the oil sands surface mining projects, and as a province it's time to admit to the people oil sands regions have been forsaken (Stop and Tell Our Politicians, Testimony at Oil Sands Consultations, Bonnyville, April 10, 2007).

I have been up to Fort McMurray, I've been to the Buffalo—I've seen it. I've looked at the replanted forests. I have forestry training, and it doesn't look very real to me. It's not reclamation (Green Communities Edmonton, Testimony at Oil Sands Consultations, Edmonton, April 4, 2007).

Members of the public who have testified at the hearings appear not to have been fooled by corporate magicianship. Members of the public insist on a definition of reclamation that embodies ecological integrity.

Given what we currently know about the slow rate of recovery (if at all) of destroyed or disturbed areas of boreal forest.... there is a realistic possibility that development of the total number of anticipated individual oil sands projects, even if staggered in terms of start and end dates, will result in the destruction or fragmentation of a significant part of the boreal forest region of Alberta.... Unless the government allows scientists to collect the data that are necessary for the development of successful reclamation strategies, we can only expect to be marginally successful in the clean-up stage and Albertans will likely be left with massive-scale loss of boreal forest.... My vision for the development of the oil sands is very much the result of the many unknowns and uncertainties that currently exist.... as well as the questionable reality of reclamation (Oil Sands Consultations, Edmonton, October 4, 2006).

So, oil sands development is accelerating as I'm writing this, but as far as I can tell, no one really knows for sure what reclamation really means; isn't this a fairly important point? How can anybody make a decision as to whether more oil sands should be developed if we don't even know what kinds of landscapes we will be leaving our future generations?.... This seems to be another example of government and industry assuming either (a) some kind of future technology or expertise will take care of present problems, or (b) that citizens are so far removed from the impact that no matter what the end-landscape looks like, the economic benefits will have made it worth it. It boggles my mind that after over a century of hindsight, this attitude can still exist (Written submission to Oil Sands Consultations, n.d.).

Testimonials also suggest that concerned citizens are not inclined to defer to corporate expertise, but rely on their own faculties for evaluation, referring not only to official documentation of reclamation but also to personal observations that are difficult to discount as "anecdotal":

> As of 2003 not one reclamation certificate was granted to any of the oil sands operators, even though some of the operators have been operating there since the 1960s. Over 30 years, and not one reclamation certificate. What does that say about reclamation processes that everyone is so excited about? (Oil Sands Consultations, Edmonton, April 4, 2007).

Finally, while industry reps draw attention to the "trees," public testimonials suggest full attention on the "forest," not only in terms of spatial scale, but also in terms of spill-over effects:

> The proposed area of development is nearly one quarter of Alberta (three times the size of Nova Scotia). I fear that this development will push us past the "tipping point," that we may not be able to reclaim this area after such an assault. Desertification, dust storms, instability of weather patterns and loss of moisture essential to agriculture in the Prairies—this is what we are toying with in our deliberations over how much free rein to give the oil and gas industry (Written submission to Oil Sands Consultations, April 11, 2007).

Aboriginal peoples living downstream from the tar sands have been a formidable critic of Alberta's sponsorship of rapid development. In February 2008, Treaties 6, 7, and 8 signed off on a common resolution calling for a stop to new oil sands approvals until a comprehensive watershed management plan approved by their Treaty communities could be implemented (AEN 2008). Storylines offered by members of Aboriginal groups certainly resonate with those of other local citizens, but are also extended with attention drawn to impacts on livelihood:

> The Métis want to be certain that.... they can exist and thrive as a people in Wood Buffalo.... Every day the land is left scarred and without reclamation efforts is a real day lost in our people's use of the lands. It adversely impacts both us and the flora & fauna (Metis Consultation Sessions, Oil Sands Consultations, Fort McMurrary, March 26, 2007).

Water

> Alberta will soon have to decide which is more valuable and important to life: water or oil (Griffiths et al. 2006).

The relationship between tar sands development and water is one that receives more regional than international attention, but the ecological ramifications are by no means less significant. To recap, three aspects of this relationship are key. First, the large volume of water consumed relative to current and future availability of supply is a concern, particularly considering climate change forecasts of increasing flow variability. Second, water pollution concerns emerge both as a result of routine operations, and, far more forebodingly, as the risk of failure of the tailing ponds. And finally, the fact that both consumption and pollution pertain to a substantial watershed stretching downstream from northeastern Alberta all the way to the Mackenzie-Beaufort Sea in the Arctic, imposes such risks onto multiple human and nonhuman populations that rely on the integrity of that watershed (Fig. 5.12).

Fig. 5.12 Tar Pond B2400738 with permission of Louis Helbig. See www.beautifuldestruction.
com and http://www.louishelbig.com/

Talk About Water

The adequacy of this policy apparatus has been a particular focus of several regional
ENGOs, which argue, for one, that Alberta's current system "is based on values
formed 100 years ago and limits these values from evolving to reflect today's knowledge
and priorities" (Griffiths et al. 2006). The recommendations of some regional ENGOs
do not depart substantially from those of industry in many respects, particularly

the emphasis on efficiency, scientific monitoring, and adaptation, but emphasize the public ownership of the resource, and – for some but not all regional ENGOs – the need to collect scientific information *before* development occurs:

> In Alberta, like elsewhere, as demand for water increases and shortages occur, management of this resource becomes increasingly important. Effective management requires a comprehensive policy framework that takes into account that water resources are public and ensures that decisions on water use are based on high-quality data and scientific knowledge. A framework must also balance current and future water demands, prevent wasteful use of the resource, weigh the relative worth of different water uses and provide adequate protection for ecosystems. So too the framework must be adaptable, allowing for changing objectives and priorities over time (Griffiths et al. 2006. Report offered as Written submission, Oil Sands Consultations, n.d.)

Other, smaller ENGOs offer a storyline rooted simultaneously in a local and global perspective:

> Water is one of our biggest concerns. As we stated during round one of these consultations, we are philosophically opposed to the removal of water from a hydrological cycle. Alberta has only 2.2 percent of the fresh water in Canada, and most of southern and central Alberta is already facing a water shortage.... Water is essential to all life on this planet and is irreplaceable (Representative of the Peace River Environmental Society, Oil Sands Consultations, Peace River, April 16, 2007).

Industry users, on the other hand, appear quite satisfied to contribute to "adaptive management," and warn that revisiting water rights would amount to broken promises with severe economic repercussions:

> Senior license rights must be respected with respect to the withdrawal of water from the Athabasca River. We are reliant on this license to increase our production capacity to a limit that has been both reviewed and approved by the Provincial regulatory process. We also note that substantial capital has been invested based on these terms.... The region's water management plan should be based on science, should be adaptable as new information becomes available, and should appropriately balance economic and environmental considerations. We are concerned that the current proposals are more reflective of risk avoidance than risk management (Syncrude Written Submission to the Oil Sands Consultations, n.d.).

At any rate, industrial representatives assure that concerns expressed are exaggerated, all the while making sure to talk in terms of *average* flows, rather than seasonal lows, which would shed a much poorer light on usage levels. The "negligible" impact is well worth the economic return (particularly when the economic value of water is considered negligible), and the rest of us are causing greater damage to the environment by wearing clothing:

> Currently, we tap into about one-fifth of one percent of the river's average annual flow. Due to production growth, we anticipate this will increase.... However, our water use efficiency remains the best in the industry. We currently import around 2 cubic metres of water per cubic metre of crude oil produced.... And to throw this into perspective, it takes about 7 cubic metres of water to produce the cotton in a pair of blue jeans (Thompson, Don, Syncrude. Presentation to the Globe Conference, Calgary, March 29, 2006).

> Findings to date suggest oil sands development is having negligible impact, if any, on the quality of local rivers and lakes.... However, by no means do these results indicate we can

rest on our laurels. Consider this—oil sand is a naturally occurring substance found all across the region.... So what we are dealing with is naturally elevated levels of hydrocarbons in the local water systems.... When you compare our cumulative economic contribution to our cumulative water withdrawal as of 2005, we have generated an economic contribution of around 7 dollars per cubic metre of imported water (Syncrude, Written Submission to the Oil Sands Consultations, n.d.).

Elected Officials representing the Opposition express concerns that often resonate with those of ENGOs; many in fact refer to reports produced by the Pembina Institute in their testimony. Their concerns appear to be handily, if not justifiably, refuted by PC MLAs, however:

Eggen: Well, considering that the main increase in water usage in this province is for large industrial projects such as the oil sands, will the minister, then, commit to a conservation system that will reduce the water consumption of large industrial projects such as the oil sands, where most of that water, in fact, is being lost?

Boutilier: In actual fact the water consumption that is being done by development in the oil sands is significant, but I also might add that their recycling, their conservation, and their off-stream storage that they have today are also excellent examples of how they have been working with our Water for Life strategy.

(Legislative Assembly, April 4, 2006)

Here is an interchange between a Member of the Opposition and a PC MLA, expressing unrelenting confidence in a bright green future:

Yankowsky: Ponds that hold the hazardous by-product, or tailings, of the oil sands are often built right next to rivers, separated from them by only a small earth dam.... What safe-guards are companies required to have in place to ensure that these tailings ponds will not fail?

Taylor: Well, first of all, let me assure the member that the tailings ponds are built only when absolutely necessary, and as the technology improves, we will need fewer and fewer of these tailings ponds. Secondly, before any tailing pond can be built, it has to receive [government] approval.... These tailings ponds are designed by experts, engineers that can do these things.

(Legislative Assembly, May 10, 2004)

MLAs expressing criticism are regularly put on the defensive by accusations from their conservative counterparts that they are antidevelopment, limiting their ability to launch an effective counter-frame. MLA Mason found himself in this position in a debate in the Assembly (on May 14 2007): "[our critique] is not, as the minister tried to suggest, an attempt on our part to say that you need to control the businesses. I said nothing of that sort whatsoever, so I didn't appreciate that very much." Not only is the existing policy apparatus sufficient, but confidence in corporate citizenship abounds (Fig. 5.13):

At present, today, as we speak, we're working with the oil and gas industry.... The next step that we foresee.... is actually setting up a protocol and a group of people, including the various industries, to look at how and what technology is available.... The oil and gas industry is being very co-operative.... They understand the necessity of developing new technologies to reduce their utilization of potable water (Taylor, Legislative Assembly, April 17, 2003).

Fig. 5.13 Tar Ponds B2402127 with permission of Louis Helbig. See www.beautifuldestruction.com and http://www.louishelbig.com/

Voices of Discontent

Members of the public are skeptical of industry goodwill, and express high levels of distrust for the government as well:

> As I see it, not only is monitoring of water quality and quantity essential, but it should be done by government agencies and not the industry (as is the case now). We can't trust a resource as important to life as water to the industries that use it (Oil Sands Consultations, Written Submission, March 31, 2007).

It would be easier to have faith in the salvation of future technologies as yet uncreated, if only our administration would strongly and honestly move toward protecting the resources we now have (Oil Sands Consultations, Written Submission, October 3, 2006).

For concerned citizens, certain forms of anecdotal and observational evidence are sufficiently compelling to demand attention.

Observations like this [anomalies found in wildlife health] need serious investigation. It would seem a very extreme coincidence if these and other observed disorders in wildlife and human populations do NOT turn out to be linked to oilsands development (Oil Sands Consultations, Written Submission, n.d.).

As an elected City Councillor (Yellowknife, NWT)—and a father—my biggest concern is the future that our children will inherit. We are downstream of the tar sands. Lower water levels are noticeable in the Northwest Territories rivers. Alberta's tar sands consume water at an incredible rate (Oil Sands Consultations, Written Submission, April 19, 2007).

And, furthermore, we have an ethical responsibility to do something about it:

The people that concern me then particularly are the First Nations people at Fort Chipewyan, Fort McKay.... they owned and stewarded the river originally, and I think it's still basically their river, and they didn't ask to be living at the bottom end of a sewer (Oil Sands Consultations, Wabasca, April 11, 2007).

Testimony by Aboriginal residents themselves offers a storyline that not only embraces local knowledge and human rights, but goes further to question the very audacity of our presumed control over such an essential life-giving force.

If I can no longer live off the land, and I am unable to eat the foods that have sustained our people for centuries, then the treaties are being broken again, because our treaty guarantees us to hunt, fish, and trap.... No matter how low the river gets, oil sands companies can keep taking the water they need. Industry's interests and making money comes first (Oil Sands Consultations, Ft Chipewyan, April 12, 2007).

Before the two plants that were here, water was good.... The people that lived around here were healthy.... Today now just about every second day you hear this one has got cancer, there's something wrong with this one, there's all kinds of different diseases now. That's with the water.... Since February, the short month, we have 17 people that have passed on in the community of Fort Chip[ewyan], and that is due to poor water that we're drinking.... I saw water going into their—their waste going into the river, directly into the river like that, that's what we have to drink and that's what's killing us all. That's why I'm saying it again.... Water is a very, very important resource. It's probably more damn important than oil.... That's what keeps us alive, and that's what nourishes our Mother Earth (Canadian Natural Resources Limited License Proceedings, Fort McMurray, September 23, 2003).

Many Aboriginal peoples espouse a perspective that reaches far beyond community-level impact, however:

Slave River is.... central to one of the largest fresh water systems in the world beginning in the mountains of British Columbia flowing down stream into the Athabasca, Peace and Slave Rivers spilling into the Great Slave Lake straight up the Mackenzie River into the Beaufort Delta of the Arctic Ocean. The Slave River Delta acts like a sink to all discharge and contamination that flow downstream from any development or activity south of it. The Dene people of Deninu Kue First Nation have used the Great Slave Land and the Slave River watershed since time immemorial and would like to continue to do so for generations to come. We need to protect the land and water from any contamination that may have impacts/effects on our Dene lifestyle (Oil Sands Consultations, Written Submissions, October 4, 2006).

This Aboriginal storyline is summarized in a written submission provided by a collective of aboriginal groups called Keepers of the Water Declaration:

> We have been taught that the land is our Mother, and the waters are her blood that sustain life for all peoples, lands and creation. We are born from our Mother Earth and we are inseparable from her. AFFIRMING THAT water is essential to life, and the right to life constitutes a fundamental human right, recognized by all countries of the world. FURTHER AFFIRMING that as Indigenous peoples, since time immemorial, have fundamental rights to live on our own lands and to survive as peoples. As a sacred trust we have been given responsibility from the Creator to ensure the integrity of all waters in our lands in all its many forms—from the aquifers deep underground, to the rich marshlands, rivers and lakes that connect and sustain our communities, to the glaciers on the high mountains, to the rains and snow that restore and replenish our Mother Earth in an unending cycle of renewal (Oil Sands Consultations, Written Submissions, September 7, 2006).

Climate

"Dirty oil" boycotting campaigns emanating from the United States and Britain have brought climate change, however reluctantly, to the center of the PC Party's political discourse. Elected officials have taken every opportunity to expound on the province's record of reductions in greenhouse gas emissions *intensity* (per gdp), as well as the small overall proportion of greenhouse gasses emitted by tar sands operations, both of which effectively counter the claims of proponents by drawing attention away from the substantive increases in total emissions evidenced over the past decade (Alberta's greenhouse gas emissions in 2008 were 41% above 1990 levels, despite a 16% decrease of emissions intensity between 1990 and 2008, Environment Canada 2010)[3] (Fig. 5.14).

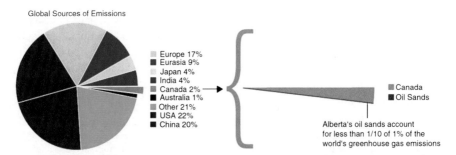

Fig. 5.14 The Alberta Government relies on graphic images such as these to dissuade concern about climate change. Used with permission of Government of Alberta. Available at: http://www. oilsands.alberta.ca/cleanenergystory.html. Accessed January 12, 2011

---

[3]All data derived from Environment Canada 2010. National Inventory Report 1990–2008: Greenhouse Gas Sources and Sinks in Canada. Ottawa: Government of Canada. Available at http://www.ec.gc.ca/ges-ghg.

Minimizing the Impact

Fearing a withdrawal of market support among Alberta's largest body of consumers, Alberta's Premier has been making numerous trips to the United States touting Alberta's affordable and secure energy and environmental stewardship. In order to protect the industry, Premier Stelmach has asked fellow Conservative, Prime Minister Steven Harper for a seat at any future North American climate change negotiations.

The PC's climate change policies to date include three main strategies. First, the province has committed revenues to support research and development into carbon capture and storage and public transit, to the tune of $2 billion each. Secondly, the Legislature has written and passed *The Climate Change and Emissions Management Amendment Act* which requires major industrial emitters to reduce their emissions intensity by 12%. The province has also established Climate Change Central, a research and outreach organization. The province has also established, as indicated in its Climate Action Plan, the goal of absolute reductions in greenhouse gas emissions of 14% by 2050, compared to 2005 levels. Analysis of the transcripts of Legislative debates as well as Premier's speeches suggests a consistent narrative in which climate change is accepted as real. The *problem* warranting attention in this narrative is not climate change itself, however, but rather defending Alberta's reputation on the international stage. The *real* climate change culprits, furthermore, are not giant energy corporations, but rather you and me:

> It's easy for some to attack Alberta's oil sands—which account for just four per cent of Canada's greenhouse gas emissions. Personal automobiles, on the other hand, create 12 per cent. And while 17 per cent of emissions produced by a barrel of oil are the result of the production and refining process—70 per cent are produced by the end user—the consumer (Stelmach, Presentation to the Canadian Urban Transit Association Annual Conference, Edmonton, May 28, 2008).

PC narratives are strikingly consistent, including characterization of industry as responsible citizen, that of environmentalists as emotional and irrational, and self-characterization as reasoned, calculated leaders. The solutions offered center on expressions of the political prudence of Alberta's existing climate change policies, while highlighting the need for accommodation of Alberta's exceptional role as a global energy supplier:

> We're coming forward with a climate change plan that starts with our province's unique circumstances. It's a plan that will deliver real, measurable reductions in greenhouse gas emissions.... while maintaining our quality of life and allowing for continued economic growth.... There's already more than enough hot air surrounding the issue of climate change. I'm not going to add to it with empty political rhetoric and targets I know we can't achieve. The plan we're announcing today will deliver real reductions, in a realistic timeframe.... without sacrificing growth or quality of life. It's a made-in-Alberta plan that fits our province.... and that will make Alberta a national and international leader in responding to climate change (Stelmach, Announcement of Alberta's Climate Change Plan, Edmonton, January 24, 2008).

The Premier has gone so far as to flip the environment/development relationship on its head, suggesting that tar sands development is necessary for generating revenues that will drive the greening of our economy. Throughout this narrative, particular frames emerge, including prescription to technological optimism, and invoking of Western cultural identity that espouses individualism and determination, all the while suggesting that placing controls on growth offers the greater threat. Here are a handful of examples:

> The challenge of climate change must be looked at in a way that will realistically attack the question, not through the ideological, rose-colored perspectives which so often cloud the view on this subject. Why not support the search for economical ways to remove carbon dioxide from the atmosphere? Just fix it. There must be a way (Backs, Legislative Assembly, March 12, 2007).

> I'm wondering what sort of a decrease in our living standards here in Alberta he [opposition MLA] would be prepared to tolerate if we were to shut down the oil sands or to cap them off and stop producing greenhouse gases. What sort of decline in living standards would be acceptable in order to achieve a zero increase in emissions? (Brown, Legislative Assembly, April 3, 2007)

Members of the opposition counter this narrative with one that characterizes the government in far more insidious terms of irresponsibility and untrustworthiness. While the voices of Liberal and New Democrat Party MLAs do not appear to have had influence on provincial climate policy to date, much of this opposition narrative resonates with the voices of citizens to be discussed further herein. Here are some examples:

> Global climate change threatens human civilization itself. The Conservative government seems unable to come to grips with its own responsibility in this matter. Uncontrolled and unplanned expansion of tar sands development not only disrupts the economy; it will soon become the source of the largest increase in CO_2 emissions in the world. The Tory government's use of emissions intensity targets is deliberately misleading. It allows total emissions of CO_2 to continue to rise dramatically while their so-called intensity drops (Mason, Legislative Assembly, March 8, 2007).

> This government continues to try to confuse Albertans by talking about greenhouse gas intensity targets while European nations and American states and even other provinces are forging ahead with actual reductions (Eggen, Legislative Assembly, March 12, 2007).

The industry narrative, which is similarly consistent, is distinct from and yet complementary to the PC narrative. As with PC Party members, industry representatives acknowledge that climate change is happening and requires a response. Industry representatives are careful to avoid the assumption of responsibility for *causing* the problem, and instead characterize themselves as part of the solution, a solution that emphasizes reliance on (future) advances in technology. Representatives also make a point of using the term "we" often, particularly when in reference to the provincial government. Use of terms like "uncertainty," "complexity," and "potential" impacts are common, as is an emphasis on society's need for energy.

> Whether man is contributing to GHG's [greenhouse gases] or nature is contributing to GHG's or indeed GHG's are contributing to global warming or not is not really the issue. The fact is we are seeing global warming and we can do our bit and the only bit we can do is attempt to minimize GHG emissions as quickly as we can. But the reason I say we can't cannot simply

Fig. 5.15 Greenpeace activists from Canada, the United States, and France place a giant banner reading "Tar Sands: Climate Crime" and block the open pit mine at the Shell Albian Sands outside of Fort McMurray, Alberta, Canada on September 15, 2009. © Greenpeace/Colin O'Connor, with permission

> give up on continuing use of fossil fuels, it's that sustainability is not just about sustainability of our environment, it's sustainability of our economy and our way of life, which like it or not requires energy in increasing quantities in the next fifty to a hundred years as driven by the developing economies.... We are a very small part of the global issue of climate change. [But] we are working very hard on reducing emissions, and in fact, are the leader in the oil and gas industry (Oil Sands Consultations, Calgary, September 28, 2006).

Voices of Discontent

Climate change is the main concern that appears to have placed the tar sands on the international radar. National and international ENGOs alike often highlight the federal government's failure to comply with Kyoto, and the specific role played by the tar sands in this failure. One of the great challenges to generating alarm about climate change is its relative lack of visualization, and environmental organizations have thus relied on metaphor in their visual as much as their textual discourse (Fig. 5.15).

ENGO spin doctors often rely on metaphorical comparisons to raise concern with the level of emissions emitted by tar sands operations, and de-emphasizing the relatively small absolute contribution (Fig. 5.16):

Fig. 5.16 Triggered by climate change concerns, boycotting has become an increasingly common strategy among nonlocal ENGOs, such as the one undertaken by members of a coalition of groups who protested outside a local BP petrol station in Tottenham Hale, London in April 2010 (BP is a major investor in the tar sands). Available at: http://sustainableharingey.blogspot.com/2010/04/report-of-tar-sands-climate-change.html. Accessed 26 January, 2011. Used with permission. www.sustainableharingey.org.uk

The Tar Sands can single-handedly prevent Canada from meeting its international obligations under the Kyoto protocol. By 2020 the tar sands are expected to release over 141 mega-tonnes of GHG—twice that produced by all the cars and trucks in Canada.... Across the country, individual Canadians are taking action to fight climate change. Most provincial governments—other than Alberta—have begun to meaningfully respond. But every step forward is undermined by ever larger greenhouse gas emissions from the Tar Sands. If we care about our planet or our future we need to STOP THE TARSANDS.[4]

Regional environmental organizations place less emphasis on climate change, and when they do, they tend to assume a moderate and collaborative position. The primary problem identified remains Alberta's (and Canada's) failure to uphold international commitments. Representatives avoid an adversarial position by offering solutions that will purportedly not disrupt development, and by playing on the same

---

[4]STOP: Stop Tar Sands Operations Permanently. Available at: http://stoptarsands.wordpress.com/. Accessed April 28, 2010.

"Alberta leadership" identity embraced by politicians. They seek buy-in by calling on corporate leadership, all the while taking care to avoid finger-pointing:

> The Alberta policy falls far short of the emission reductions that will be needed to head off dangerous climate change. WWF does not believe that a 12% intensity reduction for major emitters will achieve the significant reductions needed to prevent the adverse consequences of emission-induced climate change. A 12% absolute reduction in emissions levels would be a step in the right direction (WWF Canada, Oil Sands Consultations, Written Submissions, Calgary, April 23, 2007).

> To date we have failed in controlling greenhouse gas pollution from the oil sands. We have failed Albertans, we have failed future generations, and we have failed the global community in responding responsibly.... Oil sands are the single largest contributor to GHG emissions growth in Canada. [But] In a "can do" province like Alberta, we could achieve this vision of greenhouse-gas-neutral oil sands by the year 2020.... There are two potential ways in which the oil sands industry could take responsibility for its GHG emissions without stopping development. The first is for the industry to seek technology breakthroughs that significantly cut the GHG intensity of oil sands production. The second is for the industry to offset emissions by purchasing credits that represent genuine GHG emission reductions achieved elsewhere (Reynolds, Marlo, Pembina Institute. Oil Sands Consultations, Written Submissions, Edmonton, September 26, 2006).

> Alberta has become an energy powerhouse in Canada and has the opportunity to become a leader in Canada in control of greenhouse gas emissions in the same way that some states in the United States have moved to address this issue (NRDC, Oil Sands Consultations, Written Submissions, Bonnyville, October 3, 2006).

There are some smaller groups taking a more aggressive stand, such as the Canadian Youth Climate Coalition:

> As Albertans and Canadians increasingly witness impacts of global warming, it will become more and more difficult for our province to ignore its contributions to these environmental processes.... As a province, Alberta has a reprehensible history of discounting the necessity of the Kyoto Protocol and strict emissions reductions. This misdirection needs to be corrected and quickly changed. Our province does not exist as an island. It is interconnected with the rest of the globe, and it cannot continue to exempt itself from the battle against climate change (Oil Sands Consultations, Edmonton, April 4, 2007).

In contrast to ENGO representatives, the voices of concerned citizens offer a much deeper and more reflective narrative, one that is far more critical. This narrative is also strikingly consistent despite the fact that speakers are unaffiliated citizens. According to many citizens who testified at the provincial hearings, climate change is a catastrophe threatening civilization. The development of the tar sands, and more directly the state that endorses it, is characterized as a primary cause of climate change, and therefore the actions of tar sands proponents are directly affecting a global society of others, future generations, and the quality of life of Albertans. The Alberta government, far more than energy companies, is targeted directly for blame, and is characterized as a "drunken teenager" – arrogant, short-sighted, irresponsible – with the power to commit civilization to catastrophe (Figs. 5.17 and 5.18).

Unlike most ENGO voices, the vast majority of concerned citizens represented in our data indicated that a moratorium on development is the only feasible response, and it is our ethical responsibility to do so. A central theme that emerges in this narrative is a deep sense of culpability, and embarrassment for being associated with the tar sands in global civil society:

Fig. 5.17 Is your bank a climate bigfoot? http://climatefriendlybanking.com/ Flikr: Creative Commons Jonathan Mcintosh. http://www.flickr.com/photos/jonathanmcintosh/3404795301/

I speak as a citizen of Calgary and of Alberta, but also as a citizen of Canada and of the Earth which is, after all, our only home…. Ultimately, the issue that we face in considering the future of the oil sands in Athabasca is not economic or political, or even social. It is a moral and ethical question that strikes right at the heart of who we are (Oil Sands Consultations, Calgary, April 24, 2007).

I think we've got a responsibility to our own people and to the rest of the world frankly given the privileges that we enjoy to show some leadership in dealing with what is probably the greatest threat facing humankind of the 21st Century: the global climate crisis (Oil Sands Consultations, Calgary, September 28, 2006).

In the last year we have come to understand that global warming is an irrefutable reality. The decisions taken by this handful of people are negatively impacting the lives of people and ecosystems across the globe (Oil Sands Consultations, Written Submissions, Edmonton, April 11, 2007).

I am embarrassed and disappointed that the Province, and our country is allowing such unchecked booms and busts, such short-sighted long-term devastation to go on. It is time for leaders to wake up and take notice that the world is watching, and Alberta development policies are behaving like a drunken teenager, disrespectful of his family, which in this case are Canada and the globe (Oil Sands Consultations, Written Submissions, no location noted, April 23, 2007).

I'm finding it in some ways more and more difficult to call myself an Albertan when I travel abroad, because people around the world are very aware of the fact that our emissions are amongst the highest in the world. Is this the kind of reputation that we want to have? (Oil Sands Consultations, Edmonton, April 3, 2007).

Aboriginal people in Canada are already feeling the impacts of climate change far more acutely than are non-Aboriginal residents, due in part to their high proportion

Fig. 5.18 Anti-Tar Sands
Protest Pamphlet – Berlin,
Germany, July 2010.
The authors. Photo courtesy
of Ken Caine

of residency in the far North, where climate change is affecting the land base far
more rapidly and significantly than in the southern latitudes, and due to the fact that
a much higher proportion of Aboriginal peoples maintain a subsistence-based lifestyle in which access to local land, food and water supplies is critical. The examples
below emphasize the link between climate change and livelihood:

> We're talking about massive water allocations. As the issue of global warming, as it's affecting
> the water supplies.... and that's really an important one when we're talking about protecting
> the water.... Groundwater, surface water, glacial water, it's all going down.... So I would ask
> that you please try to consider the factor of global warming through the whole environmental
> story. It affects everything (Oil Sands Consultations, Edmonton, April 4, 2007).

> Of course climate change is something that this industry no doubt will impact in a way that's
> not healthy.... People mistake climate change for the world warming up. That's not so.
> Climate change is weather patterns, weather creations, weather ways that weather is happening.
> It changes. It changes the rules in that one day it could be cold here. Next day it could be
> hot.... And it just throws weather systems off. So that's something that we feel this project
> is going to impact and it's going to impact us. Because these kinds of things head north and
> we live north of you (Oil Sands Consultations, Fort Chipewyan, October 4, 2006).

> Climate Change and Global Warming is a reality in the north and we need to address this issue
> collectively as Canadians (Oil Sands Consultations, Fort Chipewyan, October 4, 2006).

References

Alberta Energy. (2010a). Energy Facts. http://www.energy.alberta.ca/About_Us/984.asp. Accessed 24 January 2011.

Alberta Environment. (2009). Environmental Management of Alberta's Oil Sands. http://environment. gov.ab.ca/info/library/8042.pdf. Accessed 24 January 2011.

Alberta Environmental Network (AEN) (2008). Unanimous passing of "No New Oil Sands" Resolution at the Assembly of Treaty Chiefs Meeting. http://www.aenweb.ca/node/2131. Accessed 24 January 2011.

Alberta Energy. (2009). (About Oil Sands) Facts and Statistics. http://www.energy.alberta.ca/ OilSands/791.asp. Accessed 24 January 2011.

Alberta Energy (2009b). Alberta's Leased Oil Sands Area. Available at: http://www.energy.alberta.ca/ OilSands/pdfs/OSAagreesStats_June2009vkb.pdf. Accessed June 18 2010.

Alberta Energy. (2010b). Alberta's Leased Oil Sands Area. http://www.energy.alberta.ca/ LandAccess/pdfs/OSAagreesStats_July2010.pdf. Accessed 24 January 2011.

Alberta Environment. (n.d.). Air Quality and the Oil Sands. http://environment.alberta.ca/01997. html. Accessed 24 January 2011.

All data derived from Environment Canada 2010. National Inventory Report 1990–2008: Greenhouse Gas Sources and Sinks in Canada. Ottawa: Government of Canada. Available at http://www.ec.gc.ca/ges-ghg. as footnote.

Canadian Association of Petroleum Producers (CAPP). (2010). Water – what we're doing. http:// www.canadasoilsands.ca/en/what-were-doing/water.aspx. Accessed 24 January 2011.

Environment Canada 2010. National Inventory Report 1990–2008: Greenhouse Gas Sources and Sinks in Canada. Ottawa: Government of Canada. Available at http://www.ec.gc.ca/ges-ghg.

Environmental Defence. (2010). Dirty Oil, Dirty Air: Ottawa's Broken Pollution Promise. http:// environmentaldefence.ca/reports/pdf/DirtyOilDirtyAir.pdf. Accessed 24 January 2011.

Grant, J., Dyer, S., & Woynillowicz, D. (2008). Fact or Fiction? Oil Sands Reclamation. http:// pubs.pembina.org/reports/Fact_or_Fiction-report.pdf. Accessed 24 January 2011.

Griffiths, M., Taylor, A., Woynillowicz, D. (2006). Troubled Waters, Troubling Trends: Technology and Policy Options to Reduce Water Use in Oil Sands Development in Alberta. http://pubs. pembina.org/reports/TroubledW_Full.pdf. Accessed 24 January 2011.

Hildebrand, J. (2008). Reclamation illusions in oil sands country: Lack of legislation, financial preparedness, undermine reclamation efforts. *The Parkland Post 11*(2), 1, 8. http://parklandin-stitute.ca/downloads/posts/pp_2008_springsummer.pdf. Accessed 24 January 2011.

Kelly, E.N., Short, J.W., Schindler, D.W., Hodson, P.V., Ma, M., Kwan, A.K., & Fortin, B.L. (2009). Oil sands development contributes polycyclic aromatic compounds to the Athabasca River and its tributaries. *Proceedings of the National Academy of Sciences 106*(52), 22346–22351.

Nikiforuk, A. (2008). *Tar Sands: Dirty Oil and the Future of a Continent.* Vancouver: Greystone Books.

Schneider, R., & Dyer, S. (2006). Death by a Thousand Cuts: The Impact of In Situ Oil Sands Development on Alberta's Boreal Forest. Published by Canadian Parks and Wilderness Society and the Pembina Institute. http://pubs.pembina.org/reports/1000-cuts.pdf. Accessed 24 January 2011.

Timoney, K.P. (2007). A Study of Water and Sediment Quality as related to Public Health Issues, Fort Chipewyan, Alberta. http://energy.probeinternational.org/system/files/timoney-fortchip-water-111107.pdf. Accessed 24 January 2011.

Timoney, K.P., & Lee, P. (2009). Does the Alberta tar sands industry pollute? The scientific evidence. *The Open Conservation Biology Journal 3*:65–81.

Woynillowicz, D., & Severson-Baker, C. (2006). Down to the Last Drop: The Athabasca River and the Oil Sands. Oil Sands Issues Paper no. 1. Pembina Institute. http://pubs.pembina.org/reports/ LastDrop_Mar1606c.pdf. Accessed 24 January 2011.

Chapter 6
Energy Matters

Society's short 100-year love affair with oil has been replete with regional supply concerns, heated contests over access to the globe's more substantial pools, price roller coasters, and nasty environmental disasters. Today, however, the political discourse on oil has ever so hesitantly ventured into entirely new terrain – the End of Oil. As with the close of all intimate marriages, this discourse has been replete with denial, anger, conflict, and diversion. Once handily disregarded as reactionaries, luddites, or worse, those among us who have been warning of peaks in discovery rates, production rates, and inevitably consumption rates have been heard. This is in part due to the irrefutable supportive evidence that has mounted over the 50 years that have passed since M. King Hubbert, among the most respectable of geologists of his time, first introduced the prospect among his colleagues.

Today, Peak Oil is no longer solely spoken of in hushed tones and alternative media venues, but has been tossed on the desks of the world's most prominent political and corporate leaders, and into the living rooms of literate Westerners. The Global Energy Outlook for 2009 forecast a peak in non-OPEC conventional oil production in 2010, which means that global oil demand, projected to increase 24% to 105 million barrels per day (mbd) in 2030, will need to be met from other sources.[1] The first warnings hit the presses a decade ago. Here in Canada, the *National Post's Business Magazine* ran with the headline "Energy: Are we running out of it?" in October 2001. Among the inner circles of power politics, reports began to circulate as well. Canada's Jeff Rubin, former Chief Economist at Canadian Imperial Bank of Commerce (CIBC), is among the convinced and has said as much in a number of publications. The U.S. Department of Defense Joint Forces Command warned about Peak Oil in a report in 2009, indicating official acknowledgement – and concern – for the imminence and danger of our declining ability to grow our economies on cheap oil. In February 2010, the U.K. Industry Task-Force on Peak Oil and Energy

---

[1] World Energy Outlook 2009. Accessed 16 Sept 2010 at http://www.worldenergyoutlook.org/docs/weo2009/fact_sheets_WEO_2009.pdf.

D.J. Davidson and M. Gismondi, *Challenging Legitimacy at the Precipice of Energy Calamity*, DOI 10.1007/978-1-4614-0287-9_6, © Springer Science+Business Media, LLC 2011

Security, which includes several prominent business and political leaders, issued a report: "The Oil Crunch: A Wakeup Call for the U.K. Economy," receiving much public fanfare. After having long been silent on the issue for reasons that will be discussed below, energy companies (whose business it is to know their capital holdings) are beginning to fess up. The British Petroleum Statistical Review dates a global oil production peak at 2020, at 91 mbd. A handful of jurisdictions have gone so far as to integrate such forecasts into planning. Former Swedish Prime Minister Goren Persson proclaimed Peak Oil a threat in 2005 and set out to plan for fossil fuel independence by 2020.

Until recently, naysayers have effectively planted seeds of doubt about Peak Oil by pointing to historic instances in which estimates of proven reserves have increased, adjusted in response to changing technological and economic conditions. More popular is the selective use of data, namely the adoption in discourse of "probable" and "possible" reserve sizes rather than the officially recognized "proved reserves." These estimates are based on a number of factors. First, to be considered in either of these categories by the International Energy Agency, a given deposit must be: discovered through one or more exploratory wells, recoverable using existing technology, commercially viable, and still in the ground. Given this last criteria, these numbers must indeed be understood as *estimates*. Oil reserves are thus further subdivided on the basis of what degree of certainty we can apply to these estimates on the basis of the data we have available. So, we have proved reserves, claimed to have a *90% certainty* of being accurate under existing economic and political conditions, and using existing technology. Next are the probable reserves, with a 50% confidence level of recovery, and finally, possible reserves have a 10% certainty of being produced. This latter number is often much closer to what is called Oil Initially in Place, or the total estimated amount of oil in an oil reservoir, including both producible and non-producible oil, which can be quite large. Development proponents love these large figures, but what they tend to divert attention away from is the significant difference between these estimates and the Ultimate Recoverable Oil Reserves, which represents a fraction of oil initially in place. To date, the average global Recovery Factor, or the amount ultimately recovered divided by the oil initially in place, is between 30 and 35%, and one must remember that these recovery factors represent the easy oil, so this ratio has only one direction to go – down.

More importantly, are the numbers accurate? There are good reasons to believe that even those in the *proven* reserves columns are inflated, merely due to the fact that large reserve figures are more politically and economically beneficial to the countries and corporations that control those reserves. There are certain legitimate instances in which the proportions of probable and possible reserves are shifted into the proven reserve category, such as increased precision in estimates of how much oil can ultimately be recovered, or more frequently, technological and economic developments that render once unattractive reserves attractive. The abrupt addition of the Athabasca's 176 billion barrels into the proven reserve category in 2003 by the United States Geological Survey and subsequently recognized by the International Energy Agency (IEA) is a case in point. But energy accounting over the past 30 years is marked far

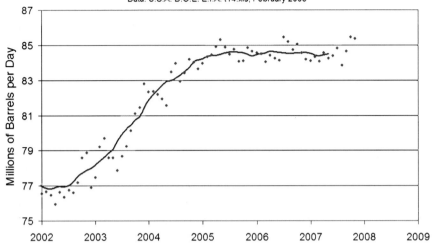

Fig. 6.1   Doug Craft. Essays on Energy Blog. Available at: www.theeternalunknown.blogspot.com. Accessed 20 Dec 2010

more often by quite questionable increases in reserve estimates. In the 1980s, several OPEC member countries showed significant one-time increases in reserve sizes – about the time OPEC decided that production quota allocations should be based on reserve size – without providing any but vague justifications. As much as 300 billion barrels of reserves officially reported by OPEC countries are considered by energy analysts to be "phantom reserves." Energy companies have similar incentives to exaggerate reserve holdings, particularly around quarterly reporting time. Royal Dutch Shell's shady accounting was revealed to the press in the early 2000s, but such practices should by no means be viewed as an isolated case.

So which numbers really matter? First, the growing gap between rates of discovery and production. What we know now, in retrospect, is that world oil discovery rates peaked in 1965, production has exceeded discovery for every year since the mid 1980s, and since about the mid-2000s global production rates have plateaued, despite periods of very high prices that, were the supply available, *should* have encouraged production growth (Fig. 6.1).

The decline in discovery rates has been attributed by some to a decline in exploration spending after a spate of consolidations around the turn of the millennium. Mergers, as always, are followed by cost cutting and profit-taking, rather than investment (it takes several years to bring a new discovery online). But, the decline is also due to the fact that all the big discoveries have already been made and new discoveries are smaller and smaller in size.

Why Worry?

All in all, even to the extent that the numbers that matter have been acknowledged, among those in positions of power such warnings have been drowned out by perseverant faith in our ability to conquer such challenges with technology or just plain luck, which of course means that very little has been done about it. How much faith is warranted? Most claims to optimism are based on the assumption that higher energy prices will drive the investments and innovation needed to seek out remaining undiscovered reserves, develop the technology to extract non-conventional resources, and contribute to a continuous trajectory of improved energy intensity per GDP. According to Jaccard (2006), for example, the problem is not the disjuncture between increasing demand and dwindling resource supply, but rather the lack of incentive structure to encourage investment in new technology that would stretch our supplies, which would consist primarily of non-conventional fossil fuels, out 500 years even assuming our historic growth trajectory in demand continues. That same technology would erase our ecological footprint as well. Industrial ecologists such as Ayres (1994) take that footprint far more seriously, although the conclusions drawn are quite similar. The premise of this school is that resources have been underpriced in the past, and the economic incentives we have relied upon to date to stimulate innovation favour short-term solutions, in many cases with long-term costs. A de-materialization of our future economies is what we need, and once again, a more appropriate incentive structure would bring it to fruition.

Meeting Energy Challenges with Oil and Gas Technologies

ExxonMobil
Taking on the world's toughest energy challenges.

Rex W. Tillerson
Chairman and CEO, Exxon Mobil Corporation
The Academy of Medicine, Engineering and Science of Texas
Austin, Texas
6 Jan 2011

It is always a pleasure to be back in Austin – and being an engineer myself, it is a special honour to address The Academy of Medicine, Engineering and Science of Texas.
This morning, I'm going to talk about the vital role energy will play in helping our economy recover, grow, and create new jobs, and how technology will be a critical enabler in meeting the world's growing energy needs.

(continued)

(continued)

In my remarks, many of the principals and challenges attendant to energy can be just as easily found in the other fields of medicine, engineering, and science. We know from centuries of history that reliable and affordable energy is essential to human progress – in Texas, throughout the United States, and indeed around the world. To sustain progress, we must continue to safely expand the world's energy supplies, improve the ways in which we consume energy sources, and address attendant environmental challenges.

As it has been throughout the history of the energy industry, technological advances will underpin solutions to all of these challenges.

Source: http://www.exxonmobil.com/Corporate/news_speeches_20110106_rwt.aspx. Accessed 20 Jan 2011.

These industrial metabolism and ecomodernist enthusiasts are quite correct in some respects. The one mainstay on which we have come to rely in historic times of resource crisis is technology. We wholeheartedly join in their calls for increased investment in research and development into efficiency and environmental improvement. The tendency, however, to assume that such measures will allow global economies to simply carry on supporting current Western lifestyles, much less improve lifestyles elsewhere, belies several basic facts about energy and economics. First, even assuming there *are* technological remedies looming on the horizon that might be of service, technology does not appear out of thin air. It requires heavy investments of capital and resources, including energy. According to the IEA (2008), the capital required to meet projected demand in 2030 is a handy $26 trillion. That represents money taken away from the rest of the economy, even assuming that $26 trillion can be generated from state coffers, and those private investors who have confidence in the profitability of such investments. Even well-established alternative technologies, such as wind and solar power, rely on fossil fuel energy sources for their manufacture. What is more, the science that enables new technologies also requires ever more resources for ever-smaller increments of new knowledge (Tainter 1988). All of which, at the very least, translate into higher commodity prices, the commodities in this case being food, water, fuel, building supplies, and pretty much anything else that must travel by ship, train, or truck to the consumer. Without requisite increases in living wages the world over, this will mean an increase in the number of families unable to meet basic needs.

Second, ecomodernists tend to conveniently gloss over the many forms of environmental degradation that are caused by our economic activities, including energy extraction, presuming such consequences can be met with the same technological innovation that will enable the energy to continue to flow. This level of optimism is problematic on several fronts. First, the irreversibility and latency effects of several forms of environmental degradation are ignored. One does not simply build soil and re-introduce extinct species with technology. The ecological impacts of resource production, moreover, *increase* in intensity as the quality of the ores declines. The ecological impact of producing one barrel of bitumen is far greater than the ecological impacts of one barrel of conventional crude, and the ecological impacts of the one-billionth barrel of bitumen will inevitably be greater than the ecological impacts of the first barrel produced. Finally, even if we were to miraculously escape or attend to these first two caveats and manage to drastically decrease our ecological footprint per barrel of oil *and* our energy per GDP ratio simultaneously, such successes are not liable to lead to reduced energy consumption or environmental degradation in a free market economy. Instead, producers and consumers alike will behave as they have historically and take advantage of such efficiencies to produce and consume more. Previous increases in energy and resource efficiency, much to the chagrin of enthusiasts, have only aided further increases in production and consumption levels (Bunker and Ciccantel 2005).

Another Spin on Our Energy Futures

These caveats describe in various ways a topic crucial to society's energy futures, but to date is much less the subject of conversation outside of petro-geology circles – something called *Net Energy*. Net Energy refers to the amount of energy available after all of the energy required to extract, transport, refine, and consume is accounted for. In a nutshell – energy efficiency enthusiasts are quite right in one key respect – we are not running out of oil. But the very solutions embedded in this narrative, in the form of investments in innovation, implicitly raise but do not problematize the crux of the matter – we are running out of *easy* (a.k.a. *cheap*) oil. And it is *easy* oil, not oil *per se*, that has driven the spectacular accumulations of wealth experienced over the past half century. There is no free lunch, but we have enjoyed a very cheap lunch ever since fossil fuels were put to use, and this boon is measured in terms of what analysts refer to as the Energy Return on Investment (EROI). The EROI effectively expresses entropy, one of the universes' basic laws of thermodynamics. The history of civilization's relationship to energy has been one of increasing EROI, right up to the point at which we began using oil as a primary energy form. The EROI for oil technically maxed out at the point at which the first barrel of oil was produced (which had an EROI of about 80:1), although until recently the EROI for conventional oil still represented a very comfortable energy profit. Conventional oil production today operates with a 10:1 EROI on average. The decline in conventional fuels and increasing reliance on non-conventional fuels represent a significant moment because we are approaching an EROI that marks what Murphy and Hall call a "Net Energy Cliff." As depicted in the graph below (Fig. 6.2), the "Cliff" is that point at which the ratio of the energy gained (*dark gray*) to the energy used (*light gray*) decreases exponentially.

Bear in mind that the math becomes questionable far sooner than the point at which the EROI reaches 1:1. This is because early EROI estimates account only for energy used at the site of production. But there is more to the story. Murphy and Hall

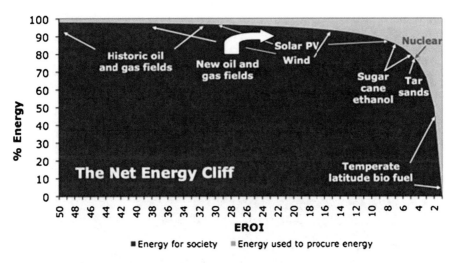

Fig. 6.2 Murphy and Hall, 2010 adapted from Mearns 2008. Used with permission

review recent analyses that assess the Energy Return on Investment *at the point of use,* which includes the energy to find, produce, refine, and transport to point of use, known as EROI$_{pou}$. As remaining deposits available to be exploited decline in quality, the effort it takes to transform those deposits into a usable energy form increases. And of course those remaining deposits are farther and farther away from population centres, which means more energy must be consumed to bring the labour and physical infrastructure to the resource, and in turn, to transport the newly exploited commodity to the population centres where it is consumed. But even this doesn't tell the whole story for oil in particular, which is used primarily for transportation. Energy analysts have thus gone one step further, to include assessment of the EROI$_{pou}$, plus the energy required to *use* the energy, which encompasses all those energy costs consumed in the manufacture of automobiles and the maintenance of our transportation infrastructure, so-called *infrastructure metabolism.* According to assessments of this extended EROI (EROI$_{ext}$), once a traditional EROI of 3:1 is reached, investments in development become questionable (Murphy and Hall 2010).

This is a purposely brief summary of such processes, which have received more detailed treatment elsewhere. But this discussion illuminates precisely why decisions made regarding the development of the Athabasca tar sands have repercussions for global society – the tar sands industrial project is the harbinger of this Net Energy Cliff. Tar sands EROI estimates range b/w 4:1 to 3:1. When placed in context of net EROI, the actual potential of the Athabasca tar sands is far less attractive. The current EROI will not be maintained, furthermore, but will follow a continuous declining trajectory, as the quality of the remaining resource diminishes (requiring ever deeper drilling shafts, greater amounts of solvent, etc.).

What is more, the predominant type of energy used as an input to tar sands production just so happens to be the same form of energy the majority of Canadians and many Americans use to heat their homes. Approximately one quarter of all energy

Fig. 6.3 The average commute distance in Toronto, Canada's largest city, is 6 miles. In the U.K., the distance is 8.5 miles, noted as "the longest in Europe" (news.bbc.co.uk/2/hi/uk_news/3085647. stm). The average one-way commuting distance among Americans, according to a NYT poll, is 26 miles (abcnews.go.com/Technology/Traffic/story?id=485098). Photo source: http://upload.wikimedia. org/wikipedia/commons/5/55/San_Jose_Freeway_Interchange.jpg. Accessed Jan 20 2010

consumed by Canadians is natural gas, estimated in 2009 at about 223 million m³/day (7.8 Bcf/d), or about 54% of Canadian production, the remainder being used by industry and shipped south of the border (NEB 2009). (At the moment, a negligible amount of natural gas is imported to Canada). Now, natural gas contributes roughly 11% of the energy content in a barrel of tar sands oil (which will only increase with in situ recovery growth),[2] and as of 2010, Alberta produced 1.5 mbd of bitumen.

---

[2]"Predictions for Canada's Natural Gas Production." Posted by Benk 2008. *The Oil Drum*, http:// canada.theoildrum.com/node/4073. Accessed 7 Nov 2010.

Surface mining has accounted for about 55% and in situ for about 45% of the total bitumen production for the past couple years (ERCB 2009). If all goes well, this figure will rise to anywhere from 3 to 5 mbd in the coming decade. Even assuming optimistically that efficiency improvements can keep up with the declining quality of the ore, they are extremely unlikely to keep up with this increase in the volume of production.

So, where will the natural gas come from? This is not clear (Soderbergh 2005), but certainly not Alberta. Even though the number of producing gas wells in the province has increased significantly year over year for the past decade, gas production reached its peak in 2001. Our next best bet is the North. But even the total anticipated gas production from Mackenzie and Arctic sources (estimated to reach 1.5 billion cubic feet per day total by 2022), both of which will be exceedingly expensive to bring online, will not be enough to sustain even tar sands production, much less offer heating fuel to Alberta households. Even Petro-Canada admits the costs of boosting natural gas production in the region have become prohibitive, and corporations are taking their investment dollars outside the region to more lucrative fields (Haggett 2003). Prospects for non-conventional gas, from shale, for example, are not great, and have a host of environmental problems themselves. Importing from gas-wealthy places like Russia is a technological and financial non-starter. Expected demand by the tar sands, combined with the fact that Canadian negotiators of the North American Free Trade Agreement in all their wisdom promised Americans that we would continue to send the same proportion of our natural gas (and oil) south every year, translates into a rather foreboding picture for Canadians in winter. Even at current consumption rates, all of Canada has only 8.7 years of proven natural gas supplies remaining, and some citizens are justifiably a bit concerned about the fact that the energy source we have come to rely on to heat our homes is being sucked up by SAGD wells in northern Alberta. Some operators are employing co-generation techniques, which make use of certain by-products of the extraction process as feedstock, to supply some of the energy requirements, but for the most part the energy requirements are supplied by natural gas, as the most efficient (and cheapest) source of heat generation available in these parts.

Why does tar sands production suck up so much energy anyway? Let's look at the process required to turn sludge into a commodity. The tar deposits are made up of mostly sand, silt, and clay. The amount of bitumen is typically around 10%.[3] Some of the deposit is relatively close to the surface (i.e., within 100 m (Isaacs 2005)), and this ore is extracted the way most such minerals are, by digging it out. These surface mining operations have predominated the industry (and the media) to date, mainly because it is the easiest to access and the extraction technology is well established; not so for the remaining 93% of the deposit too deep to mine, which we will come back to. But first, the mining process. The boreal ecosystem must be removed, euphemistically referred to in the industry as the "overburden,"—as if it didn't belong there

---

[3]From "Oil Sands Consultation Multistakeholder Committee Interim Report Appendix IV: Fact Sheets."

Fig. 6.4 The truck on display in Washington was about ¼ the size of actual vehicles used in oil sands mining. Flikr Creative Commons – Photographer Kendrickhang http://www.flickr.com/photos/kjh7r/180315120/in/set-72157594185213431

in the first place—which is typically stored onsite. Then the ore itself is dug out. According to one estimate, on average some four tonnes of "overburden" above the deposit and two tonnes of oil sands material itself are removed for each barrel of synthetic crude oil produced (Woynillowicz et al. 2005). The sand mixture is then thrown into what essentially looks like a giant washing machine with very hot water, to separate the sand from the bitumen. This enormous movement of material not surprisingly requires an equally huge amount of energy. And some pretty gargantuan equipment is needed to move it. The larger the truck, the more efficient it is, which drives the constant increase in truck size from a few tonnes in the earliest operations, to the 400-t capacity trucks used today (picture a truck 15 m long by 7 m tall, with 4-m tall tires and 40% heavier than a Boeing 747 airplane) (ibid.) (Fig. 6.4). One of the spin-off benefits of this efficiency drive is the Western cultural attraction for all things big, particularly trucks. As mentioned in an earlier chapter, Alberta displayed its mega-toys in Washington during the Smithsonian Folklife Festival in 2006, ostensibly to portray Alberta's cultural heritage, not their staples economy.

But only around 7% of the total deposit can be accessed via surface mining.[4] The remainder is too deep, and thus must be "liquefied" before it comes out of the ground. Various means of deep-well extraction are still being toyed with, but the ones used

---

[4]Ibid.

predominantly today are called in situ and steam-assisted gravity drainage (SAGD), the differences between which are not critical to the discussion here. While mining recovers about 90% of the bitumen deposit, in situ and SAGD technologies are only capable of recovering around 50% of the deeper deposits, which is one of the reasons why there is such a discrepancy between estimates of the oil in place and that which is recoverable. As the name implies, a water source, usually groundwater (some potable, some not), is heated to make steam, which is injected into the deposit and superheats the tar so that it can travel up the well pipe. Solvents can be used as well, which, as proponents are quick to point out, mean less water use, but then it also means, well, more solvent use. Given the heavy energy and infrastructure require-ments, it is far more efficient to do SAGD and in situ at a large scale, so these opera-tions, while appearing to involve a smaller footprint on the land base than a surface mine, nonetheless render very large tracts of land unavailable for ecological services. A single well pad, which houses up to 20 wells, requires complete clearance of vegetation over several hectares. One operation may encompass 25 such pads in a concentrated area, and the land between the well pads is criss-crossed with pipelines which carry water, steam, and the bitumen, as well as service roads.

Next, in much simplified terms, whether mined or sucked out of wells, a number of processing steps must take place before anyone puts this stuff into their gas tanks. First, solvents are added to make the tar viscous enough to travel by pipeline from the extrac-tion site to upgraders 500 km south in the Edmonton area, or further south to upgraders in the U.S. (about 65% of upgrading occurs in Alberta today, the rest south of the border). Even with the addition of solvents, it takes 3 days for a molecule of tar to travel by pipeline from Fort McMurray to Edmonton.[5] The tar requires two levels of upgrading. First, excess carbon must be removed, and currently the best means of doing this is by coking or super-heating the bitumen to at least 500°C, which breaks the molecules apart. The removed coke becomes a waste product. Then heat is introduced again, this time in combination with high pressure, to remove the high content of nitrogen and sulphur. At this point, the fuel is *still* not quite ready for end use, but is sent to refineries for further processing into gasoline. And of course the fuel must travel from the point of extraction, to the upgraders, to the refiners, and then once again to the retailers – all by pipeline, except for the final stage, which still travels by truck.

All of which requires energy, and lots of it, and this is the real, albeit rarely told, story. According to the National Energy Board, a conservative information source if there ever was one, as of 2004 mining requires an average of 250 cubic feet of natural gas/barrel of bitumen extracted. In situ operations, on the other hand, demand a whopping 1,000 cubic feet of natural gas per barrel (Fig. 6.5). Upgrading sucks another 500 cubic feet per barrel (Woynillowicz et al. 2005) and these numbers of course do not include the multiple additional energy requirements discussed by Murphy and Hall. The math is pretty simple from here. If we assume current pro-duction rates of 1.5 mbd, and also assume that 55% of this is mined, the tar sands

---

[5]From "Oil Sands Consultation Multistakeholder Committee Interim Report Appendix IV: Fact Sheets

Fig. 6.5 A typical SAGD operation, this one at Pelican Lake, Alberta. B2260094 with permission of Louis Helbig. See www. beautifuldestruction.com and http://www.louishelbig.com

extraction and upgrading currently consumes over 1.6 billion cubic feet of natural gas every day. If we assume further that the projections of 5 mbd by 2020 are actually achieved, and that by this time, let's say 80% of the product is extracted using in situ or SAGD, then using current technologies the tar sands will consume 6.6 billion cubic feet of natural gas per day. As mentioned, today's operations are also exploiting the best reservoirs; as these become depleted, operators will need to move into lower quality reservoirs that will be associated with lower recovery rates, which means more energy inputs needed per barrel of output, amounting to ever-mounting energy requirements to sustain projected production volumes that technological innovations in efficiency would be hard-pressed to match.

In a nutshell, we are running out of cheap oil, which is one key reason deposits such as the tar sands, until recently not considered by investors to be worth the effort, are being exploited (the other key reason is, as discussed, the steadfast marketing campaign by the provincial and federal governments). The disintegration of our relationship with oil will inevitably be slow and bumpy, however, allowing for lots of room for discursive interpretation of this relationship, interpretations that have the potential to prevent a pro-active transition to a post-oil society, a transition absolutely necessary if we are to avoid large-scale social and environmental calamity.

Energy Politics Today

Energy, and oil specifically, has been at the centre of politics and foreign relations since World War I, and particularly since World War II. It is within this legacy conflict frame in which energy security has been discussed and continues to be raised in discussions of tar sands development. And of course supply shortages will intensify international tensions. Environmental disruption has entered into political discussions about energy as well. In fact, many trace the origins of the modern environmental movement to a blow-out at an oil rig off the shores of Santa Barbara in 1969. The majority of the energy-environment discourse in the decades since that event has pertained to energy conservation and end-user pollution from cars, however, and it is only recently that the environmental disruption due to production processes has become a topic of note. As far as supply concerns go, there would still appear to be a rather large chasm between the accounting reports mentioned above and the political discourses that take place in contemporary settings.

As such, this chapter highlights what is *not* being discussed to a greater extent than what is. But first, what *is* being discussed on the topic of energy in tar sands discourses? Security, in many guises, is the centre point of discourses from nearly all parties engaged. Peak oil, consumption, and EROI rarely make appearances. As mentioned above, the tar sands' energy cliff will likely materialize as mineable sands become depleted and operations become increasingly reliant on deep-well production. But our data suggest that this energy cliff is not on the perceptual horizon of either proponents or critics of tar sands development. When in situ or SAGD are discussed at all, water use and land fragmentation are mentioned frequently, future technological innovation once in awhile. All told, the nature of this discourse, both what is said and what is not, provide further legitimation for tar sands development, first by treating demand increases as a natural and inevitable phenomenon. When declines in global reserves are acknowledged, they are used to highlight the elevated prominence of the immense pool of Athabasca reserves as a last bastion of sorts. Energy security is often raised by proponents as an asset, as Canadian resources are held up as the safe alternative, while critics are mainly concerned about the large proportion of "our" oil we are obliged to export to the U.S., according to NAFTA. We show how this discourse unfolds below.

The rationale provided for tar sands development often begins with a narrative that places Alberta at the epicentre of an "energy hungry world." Embedded in this narrative is the unquestioned assumption that demand increase is a natural and inevitable trajectory that simply must be met, once in a while including an acknowledgement that global oil reserves are declining. This narrative is a particular favourite of industry representatives. Shell Canada states on its web page[6] that:

> The global population has more than doubled since 1950, and is set to increase by 40 per cent by 2050; demand for energy is growing rapidly, as countries including China and India enter the most energy-intensive phase of economic development; supplies of easily accessible oil and natural gas are unlikely to keep up with demand after 2015. The world will have to use energy more efficiently and increase its use of other sources of energy. This means more renewables like solar, wind and biofuels, more nuclear energy, more coal and more oil and natural gas from difficult-to-reach locations or unconventional sources like oil sands.

Suncor offers the same necessity narrative, stating on its web page[7]:

> As conventional basins decline, developing the oil sands is key to keeping up with the growing global demand for cost-effective and secure energy.

This runaway demand is a further justification for doing away with pipedreams of solar panels and wind turbines, and accepting our fossil fuel reality, as noted by a representative from Suncor at the recent Provincial consultations:

> Another hard truth emerges. As much as we all believe in conservation and renewable energy ... the development of hydrocarbon resources will, for the foreseeable future, remain an essential fact of life (Williams, Steve, Suncor Energy Inc. Oil Sands Consultations, 19 Sept 2006).

> The firm reality is that global demand for energy isn't going away ... The supply has got to come from somewhere, and rest assured somebody will fill it. (Anderson, Brad, Alberta Chamber of Resources. Oil Sands Consultations, Edmonton, 4 April 2007).

And fill it as fast as possible, so the story goes, as explained by this industry representative:

> This year, we will complete a major expansion to our facilities that will increase our production by almost 50 percent. So why the increased activity? It's about supply and demand. Canadian conventional oil production is on the decline even though demand is on the rise. Oil sands could meet a large part of North American demand, while providing a stable and reliable source of energy for many years to come. (Thompson, Don, Syncrude Canada Ltd. Speech to The Globe Conference, Vancouver, 29 March 2006).

The "inevitability" of demand growth becomes an "inevitability" in production growth to meet demand. Considering that supply and demand are in a dialectic relationship, this trope becomes a self-fulfilling prophecy, since increases in supply availability lead to lowered prices, which in turn lead to increases in consumption. Another proponent even implied as much, lamenting the fact that Alberta bitumen had "become undervalued against competition because of oversupply in the markets which they currently have access,"[8] but this implication was used to support the next

---

[6]http://www.shell.ca/home/content/can-en/aboutshell/energy_challenge/. Accessed 21 Dec 2010.

[7]http://sustainability.suncor.com/2009/en/responsible/979.aspx. Accessed 21 Dec 2020.

[8]McInnis, David, Canadian Energy Pipeline Association, Oil Sands Consultations, Calgary, 27 Sept 2006.

"inevitable" development step, expansion of infrastructure to support the increase in supply, a lament shared by other industry supporters:

> Comparing the forecast growth of crude oil exports with the forecast of the spare pipeline capacity for exports, crude oil pipeline capacity leaving Western Canada will be very light by 2007 to 2008. And is expected to fall well short of requirements by 2010. This shortfall is forecast to exceed a half million barrels per day by 2012 ... . And further complicating matters is that Canadian refineries are forecast to increase their crude runs by about 45 thousand barrels per day by 2010. Clearly this will only take up a very small amount of the anticipated increase in crude production. (McInnis, David, Canadian Energy Pipeline Association, Oil Sands Consultations, Calgary, 27 Sept 2006).

In a nutshell, "inevitable" increases in demand translate into an "inevitable" need to expand production, which leads to an equally inevitable need to supply the necessary infrastructure to move the product to those making the demand. Nowhere in this discourse is the fact that there is nothing whatsoever inevitable about future demand – demand is shaped by all manner of political, economic, and technological conditions, *and* it is also fundamentally shaped by *supply.*

The Discourse of Immensity

This conversation would be mere trivia without another key actor in the narrative, the resource itself. In an energy hungry world, Alberta's tar sands are offered up as the solution, placing the future of world society squarely on the shoulders of Albertans. In this narrative, of course, size matters, and this is where the discourse of immensity comes in. Or returns, since it first emerges during the age of discovery, as discussed in Chapter 3, when the tar sands were first characterized in awe-inspiring terms. We see it in McConnell's geological survey estimates that introduced the image of the World's Largest Oilfield. Such imaginaries were quickly embraced by Alberta's early European settlers. An advertisement in the *Red Deer News* dating back to 1925 sponsored by Aurora Oilfields quotes a well-known geologist, who described the tar sands as "without question the largest exposure of oil in the known world."[9]

As we fast forward to the 1970s, when commercialization finally began to look realistic, we see the discourse of immensity re-engaged in the popular press, as illustrated by the words of author Joe Fitzgerald, in his 1978 book, *Black Gold with Grit* (1978: xiii):

> Few people realize that in the oil sands deposits of northern Alberta, in western Canada, there is enough heavy oil, mixed with sand, to pave a four-lane super highway the entire 250,000 odd miles (400,000-odd km) to the moon, with ample to spare for approaches and exit ramps. With this resource, Canada has the largest single deposit of liquid oil in the world.

Today's rhetoricians could easily be mistaken for yesterday's, although current discourse has the further advantage of official reserve estimates (officially recognized

---

[9]*Red Deer News* 3 June 1925 p.4

as 176 billion barrels). One reporter referred to "the vast ocean of tar-like goo in the northern part of the province [and] By most estimates, there is more oil in the so-called 'tar sands' than there is in all of Saudi Arabia, or about 300 billion barrels that is recoverable using existing technology" (Ingram 2001). Paul Chastko's book (2004: xiii) makes use of such estimates for the purpose of showcasing Alberta in global comparative light. Note the seemingly casual but, in actuality, very strategic employment of two very different categories – proven reserves to describe Iraq's and Saudi's seeming mediocre supplies, and Alberta's *probable* reserves, a number fantastically greater than the proven reserve estimate of 176 billion barrels:

> It holds more oil than Iraq's proven reserves of 112 billion barrels. It even holds more oil than Saudi Arabia's 250 billion barrels. In fact, Alberta's oil sands deposit contains between 1.75 and 2.5 *trillion* barrels of oil—approximately 200 billion barrels of which are recoverable with current technology. That is enough oil to supply all of Canada's petroleum needs for the next 475 years. In fact that is enough proven reserves to supply all of North America's petroleum needs for the next forty-seven years—without using a single drop of oil from another source.

Not surprisingly, industry proponents have capitalized on this discourse of immensity, and several make a point of joining Chastko in his tendency to emphasize probable reserve figures (Scott and Lewis 2004:1):

> The oil sands resource is almost unfathomable with a bitumen volume in place of approximately 1.6 trillion barrels and some 175 billion barrels of recoverable reserves in the ground. Today, oil sands companies produce about 1,000,000 barrels/day, enough to satisfy 62 per cent of Canada's energy needs. By 2015, oil sands production will account for three-quarters of all western Canadian production.

Albertans, by extension, should feel proud of their tar, but also need to embrace their responsibility to produce this global treasure. Proponents work this narrative first by reminding Albertans of their global stature:

> As Albertans, we have the good fortune of being at the centre of one of the world's greatest natural resource treasure troves … and people want this stuff too. (Anderson, Brad, Alberta Chamber of Resources. Oil Sands Consultations, Edmonton, 4 April 2007)

… and secondly, by translating fortune into responsibility, as emphasized by this MLA:

> I think it is patently naive to think that Alberta is just going to curtail their production in oil. We have over 80 per cent of North America's oil reserves right here in our province. I just think it's absolutely unreasonable. (Oberle, Alberta Legislative Assembly, 11 April 2007).

While proponents often offer future demand projections as rationale for rapid development of Alberta's treasure trove, concerned citizens offer a powerful counter-frame, referring to that treasure trove itself as justification for a more cautious pace, albeit one that has not yet unseated that of proponents:

> What is the rush to expand the projects? I don't understand the claim that the economy will collapse if we don't keep expanding and the developers will go elsewhere. We'll still have the oil and we're still going to need it far into the future. (Oil Sands Consultations, Edmonton, 3 April 2007)

Tar Sands Means Canadian Energy Security

Chastko's reference to meeting Canada's energy needs is quite handy in fomenting support in and of itself. As we will elaborate shortly, this trope is as deceiving as is the play on numbers illustrated above. But, banking on an assumed lack of awareness among voters and consumers, it has been adopted by numerous others, not just in verbal exchanges but also in official government documents. And why not? Who could argue, in the current era of high energy prices and turbulence in the Middle East, with politicians and CEOs expressing concern about our ability to keep the lights on and cars running? The opening statement in a draft of the *Mineable Oil Sands Strategy (MOSS)* reads simply, "Alberta's oil sands are essential to a secure North American energy supply,"[10] a "fact" used to justify a proposed land management plan that gives tar sands development top priority above any other concerns or land uses in a designated development area of approximately 2,800 square kilometres. Around the same time frame, as preface to the provincial Oil Sands Consultations, a Terms of Reference was developed to guide the process, which states, once again, that "the oil sands area is key to Alberta and Canada's energy security, with $80 billion worth of projects already announced."[11] This and similar prefaces to the proceedings were very effective in establishing the parameters of debate. In short, development will not be subject to questioning. In other venues, this same narrative has been used to justify this country's failure to comply with Kyoto, as Ralph Klein did in a speech to the Empire Club of Canada back in 2002, in the heat of Kyoto negotiations. In classic Klein prose, the former Premier told some of Toronto's most powerful members of the business elite "Those sands contain more oil than all of Saudi Arabia. ... Alberta's oil sands are more than sand; they are the promise of energy security for Canada for decades to come." Which is why, says King Ralph, we need a Made-in-Canada approach to climate change that won't put the brakes on extraction.[12]

Contradictions Revealed

This Canadian energy security narrative was not particularly difficult to unseat under the blatant circumstances of the Canadian energy political economy, in which 75% of tar sands production is exported to the United States. Canada is foremost a staples state that has for a century served as a source of raw materials to the United States, a role formalized in the North American Free Trade Agreement. Oddly enough, the very same actors wielding this national energy security narrative had something very different to say to audiences south of the border, where proponents

---

[10] http://www.energy.alberta.ca/OilSands/pdfs/MOSS_Policy2005.pdf. Accessed 21 Dec 2010.

[11] Terms of Reference - Oil Sands Consultation Group, 20 Dec 2005.

[12] 23 October 2002. Presented to A joint meeting of The Empire Club of Canada and The Canadian Club of Toronto.

were eagerly seeking investors and markets. Alberta's aggressive marketing campaign took full force under Ralph Klein's Premiership, his crowning achievement being a personal meeting with then-Vice President Dick Cheney just months before Sept 2011. While recordings of this meeting are not publicly available, in an interview with a news reporter, Klein said he "opened [Cheney's] eyes to the vast potential reserves buried in the Alberta tar sands and to the province's willingness to quench the U.S. thirst for secure energy," assuring Cheney that "we have energy to burn."[13] With Canadian industry joining in on the marketing efforts, the idea took root, with U.S. agents such as Spencer Abraham, Energy Secretary, singing the praises of Canadian tar as an alternative to Middle East oil.[14] Shortly thereafter, the U.S. Geological Survey upgraded the reported Canadian reserve from five billion barrels to 181 billion (soon corrected to 176 billion).

Development of the Alberta tar sands, in short, is all about America's security, not Canada's. One MLA perhaps did not recognize the weight of his clear admission in this oratory in the Legislature in 2005:

> Our good friends and neighbours to the south, the United States, have been in discussions for some years, actually, on an energy bill that would really look at their energy security. Alberta figures prominently in that role, given that we are the largest source of both oil and gas to the United States … they have listened to Alberta. (Melchin, Alberta Legislative Assembly, 5 May 2005).

This particular MLA was not the only one to admit as much; his colleague did the same a year earlier. In reference to a meeting between Canada's Prime Minister and the U.S. President, another MLA stated: "the goal was to reinforce the United States' understanding and appreciation of Alberta as a crucial energy security source and particularly the contributions that could be made by increased U.S. investment in Alberta's oil sands."[15]

More recently, while on a trip to Washington, the current Premier re-asserted the Alberta campaign, aided by the weight of several years of Middle East turmoil:

> If the last few years have taught us anything, it's that real energy security requires secure, reliable, affordable energy. That's what Alberta offers—and it's right here in North America within a politically stable, U.S.-friendly, and business-oriented jurisdiction. Alberta is positioned to play a vital role in U.S. energy security. That's a role we want to play. Alberta is second only to Saudi Arabia in global oil reserves … [and] the only non-OPEC oil producer with the potential to substantially increase energy production in the short-term. Those two facts put Alberta at the epicenter of a new world order in energy. (Stelmach. Alberta Enterprise Group Energy Forum, Washington, DC, 16 Jan 2008).

This targeted marketing campaign makes perfect strategic sense. As it happens, the U.S. represents an enormous percentage of global consumption, so there has

---

[13]McKenna B 2001. "'We have energy to burn,' Klein says." *The Globe and Mail,* National News Section, June 15.

[14]Cattaneo, C. 2003. *National Post.* FP03.

[15]VanderBurg, Alberta Legislative Assembly, 4 May 2004.

been limited incentive to seek other markets anyway. But a more formidable reason is purely geographic. The Athabasca is land-locked, and thus the synthetic fuel is non-fungible – since it must travel by pipeline, it does not get sold "on the open market," it gets sold to whoever is on the other end of the pipe. As a result, Alberta is highly unlikely to ever become the centre of a "new world order" in energy *markets*, as Stelmach would have us believe, although one could argue that, with investments coming from corporations across the globe – both private and state-owned – Alberta could be described as the centre of a New World Order of energy *finance*.

The export pipeline infrastructure that has been in place for decades travels in one direction only: from Alberta's processing centres to upgraders and refineries in the U.S., primarily in the Midwest. Other prospects are foreclosed due to the extraordinarily high costs associated with pipeline construction. At the moment, when concerns have been raised that existing pipeline infrastructure will not be sufficient to carry expected increases in production, the potential for reconsideration of target markets has been raised. There have been several recent attempts to develop pipelines to the western coast of British Columbia for marine transport to gain access to markets in Europe, Asia, and anywhere else a tanker can travel, but to date these efforts have not had much purchasing power. Traversing the difficult topography makes the endeavour extraordinarily expensive, deepwater shipping terminals themselves must be built, and routes considered so far encroach on the reserves of First Nations, which have justifiably put up a vehement fight. Interestingly, an easterly route toward Canadian markets in Ontario and Quebec has not even been given serious consideration.

Market (Ir)rationalities

As of 2009, Canada produced 4% of the world's supply of crude oil, about 152 MT, or 3.3 mbd. That year, the country consumed 2.15 mbd, but 1.16 million of those barrels needed to be *imported*, because 2.42 mbd produced in Canada left the country, almost entirely headed to the United States. The vast majority of the oil consumed in Canada comes not from the tar sands, but from conventional sources, primarily in the east, where much of the population is located. So what becomes of the oil produced in the tar sands? Just follow the pipelines – they all point south, contributing to American, not Canadian, energy security. Even there, the contribution of Athabasca tar does not relieve the U.S. of dependence on Middle East sources to any significant extent. If current projections of future production and demand pan out, the tar sands production of 2.7 mbd in 2015 would only meet 11% of the United States' projected demand. Looking to 2030, the 5 mbd of projected tar sands production would only fulfil 16% of U.S. demand (Sources: Toulin 2001; CIA World Fact Book https://www.cia.gov/library/publications/the-world-fact-book/geos/ca.html, accessed 5 Sept 2010.

It is this "Alberta for sale" mindset that has many critics calling into question just how this spectacular sale of Canadian resources does much of anything for our own energy security. As stated by one citizen who testified at the Oil Sands Consultations,

> I don't think we're increasing our security—obviously we aren't when we gave away our energy security in the free trade agreement in 1988. So we're not increasing Albertan's security in energy or electricity or water or food security or any other thing. (Oil Sands Consultations, Calgary, 28 Sept 2006).

More to the point, according to Opposition Party MLA Mason (who asserts his resistance, among other ways, by refusing to heed the call to refer to our treasure trove as "oil sands"):

> The entire development of the tar sands in the present circumstances is not being done, in our view, in the interests of the people of Alberta and certainly not in the long-term interests of the people of Alberta. They are being done in the interests of the foreign oil companies, the giant energy corporations, and, of course, in the interests of the George Bush administration's energy policy, which is to find a safe, secure, and relatively cheap source of energy given that their adventure in Iraq has come to a rather bad end. It's pretty clear that the invasion of Iraq was plan A for the Bush administration, and Alberta's tar sands and this government's policies around that are plan B. (Alberta Legislative Hearings, Thursday, 14 June 2007).

The formalization of Alberta's role as a staples state by NAFTA laid bare for many participants in the Oil Sands Consultations the seeming willingness on the part of the government of Alberta to give away not just national energy security, but its autonomy as well. This presented a formidable legitimacy challenge for the state: either the members of the Multistakeholder Committee (MSC) overseeing the Oil Sands Consultations are woefully naïve, or the Consultations themselves are nothing but a symbolic exercise intended to assuage those legitimacy challengers. This was a particular issue for the Parkland Institute:

> Under the provisions of the North American Free Trade Agreement, every barrel shipped across the border permanently affects our ability to supply for the energy needs of Albertans and Canadians … A government serious about managing oil sands development must begin by taking back control of its ability to manage trade. (Acuña, Ricardo, Parkland Institute. Oil Sands Consultations, Edmonton, 4 April 2007).

> Indeed, Alberta has plenty of oil, more than enough to meet Eastern Canadians' needs, and export surpluses. But Alberta cannot supply Eastern Canada, even if a crisis hit and they were freezing in the dark, because NAFTA reserves Alberta's oil for Americans,' not Canadians,' security of supply. (Laxer, Gordon, Parkland Institute. Written Submission, Oil Sands Consultations, 25 Sept 2006).

> The major problem with Canada and Alberta is that neither has an Energy Security Policy. NAFTA says that we cannot cut back shipments of oil or gas from the volume we are sending them now, even if we need the products here. Add to that the fact that we do not stockpile any of our production for emergencies, as the Americans do, and I have to wonder where our politicians' loyalty really lies. While the members of our government and the MSC might think they can make these kinds of decisions, I would advise them to read the trade agreements Alberta has signed or is obligated to comply with as a Canadian province. (Unaffiliated citizen, Written Submission, Oil Sands Consultations, 31 March 2007).

We Are (North) Americans

While allusions to Canadian energy security persist in several discourse outlets, proponents have increasingly assuaged critics by simply re-creating jurisdictional loyalties. Alberta's tar sands become "the key to *continental* energy security," a phrase used multiple times by (the current) Premier Stelmach in particular, and other proponents as well. Canadians are North Americans, after all, and our comradeship with our southern neighbours was only further enhanced on 11 Sept 2001, with an event that enabled the elevation of security frames to embrace both energy security and military security. Development of the tar sands in effect protects North Americans from terrorism. Albertans can still be proud of the fact that they live on top of tar, but we are now responsible to our North American brethren, namely U.S. commuters (Fig. 6.3):

> I'm very pleased that Alberta continues to play a ... key role in supplying energy to the entire continent. I've often said that a strong Alberta means a strong Canada. But really, a strong Alberta means a stronger North America. (Stelmach, Presentation to the 7th Annual Arctic Gas Symposium, Calgary, 1 March 2007).

> Heavy oil has an important place here in Canada's "energy province." That's a label we Albertans wear with pride, as a major energy supplier to North America ... Alberta's oil sands play a major role in Canada's economy, and in the North American energy market. (Stelmach, Presentation to the World Heavy Oil Congress, Edmonton, 10 March 2008).

> If we look overseas for supply solutions, we find ourselves relying on oil from very troubled and volatile parts of the world. And that brings us to Alberta's oil sands – a vast, secure, reliable resource that promises to provide a major part of the world's energy supply for decades to come. (Suncor Energy Inc. Oil Sands Consultations, Fort McMurray 19 Sept 2006).

Corporations to the Rescue

Every narrative needs a hero, and in our case, the heroes are the energy corporations. The real purpose of private capital, namely the generation of profit for shareholders, recedes behind the theatre curtain, and energy companies present themselves as good Samaritans. As stated on company web pages, for Syncrude, securing Canada's energy future is "what we see as our broader purpose in society." Exxon is "developing oil sands reserves to help meet the world's growing energy demand." Husky is "developing energy sources to provide for a more sustainable society." Industry representatives offering testimony in the oil sands consultations continue this characterization:

> The issue of ensuring that Canadians have the supplies of oil and oil products to heat their homes ... is an important element of our mandate. (Canadian Natural Resources Ltd. Hearings, Fort McMurray, 23 Sept 2003).

But more importantly, PC MLAs and the Premier himself do as well. This strategy has clear benefits, one in particular being justification for heavy state subsidization of tar sands development, and low royalties. We shouldn't be charging energy companies, we should be thanking them for saving us:

> We see the oil sands as a long-term source of prosperity for our province and for our country … one that, responsibly managed, will mean investment, jobs, and energy security over the long-haul. That's the future I see for this industry. It's a future that many of you, as industry leaders and investors, are helping to build. I thank you for that. (Stelmach, Address to the 6th Annual TD Securities Oil Sands Forum, Calgary, 9 July 2008).

In Situ Dreams, Untold Nightmares

If there were ever an obvious place for critical consideration to emerge of just what EROI foretells for the tar sands, and for society's relationship to energy in general, it would be in discussions of in situ extraction. And there have been a handful of admissions that allude to such consideration, although those comments could hardly be said to be critical in any way. Back in 2003, for example, when Canadian Natural Resources Limited (CNRL) was engaged in hearings prior to approval of a project expansion, one company representative acknowledged the questionable returns from in situ, but then turns around and embraces the assumption that this return rate will get *better*, rather than worse, as development proceeds:

> From an energy consumption standpoint also and from a recovery standpoint today, we believe that mining has an advantage. Mining recovers about a 90 percent recovery while in situ is around 50 percent. So we believe that in situ is going to get better as time goes by. From a resource recovery standpoint, we believe that waiting a few years has its advantage. (Canadian Natural Resources Ltd. Hearings, Fort McMurray, 17 Sept 2003).

One report by the Institute for Sustainable Energy, Environment and Economy (ISEEE) clearly acknowledges that "the process of extracting oil resources from the underlying sediments is energy-intensive, requiring continuous inputs of heat, water and steam. This is most pronounced in the in situ recovery operations."[16] The report goes on to acknowledge the "limited Natural Gas supplies generally, combined with declines in regional reserves," although consideration of this reality is limited to input cost concerns. Rather than questioning the pace of development itself, the authors assume "continued expansion of oil sands operations, widely expected to form the base of future North American oil supplies," and thus the challenge becomes consideration of alternative source fuels that will "enhance operations efficiency," one key option being co-generation or the utilization of waste heat generated from current operations. Most operations already have co-generation facilities in operation, however, and the energy inputs remain large and growing, so it is not entirely clear the extent to which expansion of co-generation facilities will change current demand trajectories for energy inputs to any significant degree.

---

[16] The Missing Link: An Evaluation of the Proposed Northern Lights Transmission Project ISEEE Occasional Papers July 2006.

In most public deliberations, however, even these acknowledgements remain below the surface, and company representatives are far more likely to sing the praises of in situ extraction in comparison to mining, during a time when images of the gaping holes in northeastern Alberta had been broadcast the world over, much to the chagrin of proponents.

> In situ reclamation must be handled differently from open pit reclamation. Through progressive reclamation of pads and the nature of the disturbance (pad clearing vs. pit mining) in situ operations have a less intrusive and smaller temporal footprint than open pit oil sands extraction. (Scott, Michael, Devon Canada Ltd., Oil Sands Consultation, Calgary, 23 April 2007).

> In situ deposits ... are developed underground with little surface disturbance. Advances in technology such as directional drilling enables companies with in situ operations to drill multiple wells—sometimes more than 20—from a single site, resulting in less land disturbance. (Canadian Assoc. of Petroleum Producers, Written submission, Oil Sands Consultation, Calgary, Oct 2006).

And not surprisingly, proponents make every effort to encourage Albertans to assume a common sense of identity and pride of ownership – of the idea if not the material benefits of in situ development:

> I believe Albertans should be excited by more than a dozen new home-grown in situ companies creating the needed technology and expanding at a more moderate pace of development than the mega-projects of recent mines. ... The in situ oil sands development has a substantially smaller environmental footprint and emissions impact than mining or upgrading operations. Our goals mirror Alberta's goals; to efficiently develop the resource. (Van Gelder, Marla, Laricina Energy Ltd., Oil Sands Consultations, 24 April 2007).

Interestingly, very few tar sands critics are paying much attention to in situ operations, despite the fact that they will necessarily overtake mining as the predominant tar sands extraction process in Alberta. A search of the websites of major environmental organizations reveals limited mention of in situ at all. Closer to home, the Pembina Institute is paying attention, although seemingly not to the whole picture. In 2005, they published a report in which they express concern that energy supply crunch will encourage nuclear power development in Alberta, or gasification of coal or oil sands residue, both of which pose even greater environmental or climate threats or both (Woynillowicz et al. 2005). A year later they published a report that explicitly raised concerns about in situ operations (Schneider and Dyer 2006), but the concerns raised are limited to in situ's much ignored ecological footprint – most certainly a justifiable concern, but by no means the only one. Pembina's writers note, for example, that

> if in situ recovery of all of Alberta's underground reserves is allowed to proceed, the area impacted will be vast—approximately 13.8 million hectares (ha), or 50 times the area of the mining zone. This equals 21% of Alberta, or a land area the size of Florida. ... The implications are startling. By even the most conservative estimate, there will be more long-term deforestation from SAGD [steam-assisted gravity drainage] development than if the entire mineable oil sands region is completely cleared. The ecological effects will be many times greater still, because the SAGD disturbances will be dispersed across a vast region.

Among members of the public who have testified at the oil sands hearings, as with environmental organizations outside the region, very few mention in situ at all and none raise questions regarding the EROI of the tar sands. Here are a handful of the exceptions:

> Also unlike conventional oil, oil sands do not release petroleum in its pure form. Operations remove bitumen (which is a heavy, carbon rich, hydrogen-poor hydrocarbon) from the mixed sand, water and clay and then the bitumen is upgraded to "syncrude" by removing carbon and adding hydrogen. Generally, it takes 28 m^3 (1,000 ft.^{3}) of natural gas and from 2.5 to 4 barrels of water to produce one barrel of bitumen. (Written submissions, Oil Sands Consultations, Calgary, 28 Sept 2006).

> We must manage the inputs. Today we're using a significant amount of natural gas, which I think has a higher more value added purpose elsewhere other than using it as fuel for the oil sands for SAGD or to create hydrogen. (Landry, Greg, Jacobs Engineering, Oil Sands Consultations, Calgary, 28 Sept 2006)

The Said and the Unsaid

There are several observations that can be made about this political discourse. The first is the contradictions embedded within the proponents' narrative. Second is the unquestioned, and at times, even unstated assumptions that enable certain alternative discourses to be swept under the rug. One of the more significant assumptions here is that tar sands development will improve security, for Albertans, Canadians, or Americans. Even full actualization of future development trajectories would not deter a significant degree of U.S. dependence on Middle East energy sources. To the contrary, this industrial project has *elevated* several forms of security threat. The first, in the energy boom town itself, consists of the multiple forms of social disorder that have historically been associated with natural resource-based communities, events that have a high likelihood of unfolding in situations when you have several young male workers awash with high paycheques, and in which many of those workers are new to the community and represent a diversity of cultural backgrounds. We discussed the challenges of boom town living further in Chapter 4, but to be brief, such conditions have a tendency to be associated with high levels of drug and alcohol abuse, prostitution, domestic violence, petty crimes, and so on, and Fort McMurray is no different. In addition, the disruptions to the regional environment are in and of themselves a significant security threat, one that has already entailed threats to the health and food supplies of Aboriginal peoples. And these risks are the outcome of industrial routine – now imagine if just one of the tar ponds were to fail, not unlike what plagued Hungary and the entire Danube river system in the summer of 2010. Perhaps most ironically, developers have created an ideal terrorist target by establishing a centralized source for the United States' energy, and one that is characterized by a decentralized physical infrastructure that is virtually impossible to defend effectively. Even home-grown amateur vigilantes in rural communities in British Columbia have managed to sabotage pipelines in that province with relative ease.

Finally, what is *not* being said is as important as what *is*. If anything, development of the tar sands is offered up by developers and critics alike as the *answer* to Peak Oil, rather than a harbinger of just what a future of non-conventional fuel reliance will look like, should we choose not to confront our addiction to oil. As stated by one industry representative at the Oil Sands Consultations:

> Conventional production in both Canada and the United States has either plateaued or is declining. The Canadian oil sands represents the future of North American oil production and energy security. (McInnis, David, Cdn Energy Pipeline Association, Oil Sands Consultations, Calgary, 27 Sept 2006).

The tar sands do indeed represent a threshold in society's relationship to energy, although it represents a potential future that is foreboding, rather than inspiring. This is the threshold at which increases in efficiency in oil production have been maximized, and now we are going in the other direction, with efficiency *declining* and relative ecological impact *increasing*. Few participants in the now-global dialogue on the tar sands, or sociologists for that matter, venture down this path of critical reflection, despite the serious implications for rationality, reason, and responsibility in our global society.

Sociological accounts of energy do exist, and such accounts clearly establish the extent to which societies have been organized in quite fundamental ways around energy sources. Changes in energy sources, as a result, present moments of social disruption. But all previous moments have been defined by *increasing* energy availability to society. In reality, since that most momentous of moments – the point at which we began to use fossil fuels for energy – these subsequent moments did not actually amount to increased energy returns on investment, but rather an increase in our ability through technology to provide energy to society in more concentrated forms. The first steam engine, despite the fact that it had a miserable energy efficiency ratio of something like 3%, represented a dramatic breakthrough based on its ability to concentrate that energy output. More to the point: as with all valued elements, the cream of the crop is used first, and the history of society's relationship with energy has been defined by a continuous trajectory down the path of declining EROI. This global EROI until very recently has nonetheless been sufficiently favourable to enable the generation of vast amounts of wealth and military strength over the course of the past 100 years. That, however, has begun to change, and we have now reached a crucial point in the history of society's intimate relationship with fossil fuels, in which the costs of oil consumption – not just in terms of social, political, and environmental costs, but of *energy* costs – call into question the rationality of continuing this relationship – a point at which the energy available to society is *decreasing*.

References

Ayres, R.U. (1994). *Industrial Metabolism: Restructuring for Sustainable Development*. Tokyo: United Nations University.

Bunker, S.G., & Ciccantell, P,S, (2005), *Globalization and the Race for Resources*. Baltimore: Johns Hopkins University.

Chastko, P. (2004). *Developing Alberta's Oil Sands from Karl Clark to Kyoto*. Calgary: University of Calgary.

Energy, Resources and Conservation Board (ERCB). (2009). *Year in Review 2008*, http://yearinreview. ercb.ca/reserves_bitumen.htm. Accessed Nov 7 2010.

Fitzgerald, J. (1978). *Black Gold with Grit.* Vancouver: Evergreen.

Haggett, S. (2003). "Nault won't oppose U.S. subsidies for Alaska pipeline," *Calgary Herald*, 22 May, D2.

Ingram, M. (2001, 29 September). Canada's oil could ease U.S. dependence on Middle East. *The Globe and Mail,* Report on Business.

International Energy Agency (IEA). (2008). *World Energy Outlook 2009.* Available at: http://www.worldenergyoutlook.org/2009.asp. Accessed Dec 21 2010.

Isaacs, E.(2005). *Canadian Oil Sands: Development and Future Outlook.* Calgary: Alberta Energy Research Institute.

Jaccard, M. (2006). *Sustainable Fossil Fuels: The Unusual Suspect in the Quest for Clean and Enduring Energy.* Cambridge, UK: Cambridge University.

Mearns, E. (2008) The Global Energy Crisis and its Role in the Pending Collapse of the Global Economy. Presentation to the Royal Society of Chemists, Aberdeen, Scotland. Available at: http://europe.theoildrum.com/node/4712. Accessed July 26 2011.

Murphy, D.J., & Hall, C.A.S. (2010). Year in Review EROI or Energy Return on (Energy) Invested. *New York Annals of Science, 1185,* 102–118.

National Energy Board. (NEB). (2009). *Canadian Energy Overview 2009—Energy Market Assessment.* http://www.neb.gc.ca. Accessed Nov 10 2010.

Schneider, R., & Dyer, S. (Eds). (2006) *Death by a Thousand Cuts: Impacts of in situ oil sands development on Alberta's boreal forest.* Edmonton: Pembina Institute.

Scott, A., & Lewis, I. (2004). *Mega Project Excellence: Preparing for Alberta's Legacy.* Edmonton: Alberta Economic Development Authority.

Soderbergh, B. (2005). *Canada's Oil Sands Resources and Its Future Impact on Global Oil Supply.* Uppsala, Sweden: Hydrocarbon Depletion Study Group. Available at: http://www.peakoil.net/AIMseminar/UU_AIM_Soderbergh.pdf. Accessed Jan 30 2011.

Tainter, J. (1988). *The Collapse of Complex Societies.* Cambridge UK: Cambridge University.

Woynillowicz, D, Severson-Baker, C., Raynolds, M. (2005). *Oil Sands Fever: The Environmental Implications of Canada's Oil Sands Rush.* Edmonton: Pembina Institute.

Chapter 7
Lessons from the Study

The consequences to the ecological foundations of human (and non-human) life from climate change, losses in biological diversity, trans-boundary waste movements, rising sea levels, changes in weather and food production, and so on are of a magnitude not previously experienced and pose not just a threat but a certainty of changes to our collective quality of life. These processes, and the complex global systems that they are altering, have had significant consequences for politics and raise questions about the utility of certain theoretical concepts formulated under very different circumstances. The tar sands enterprise comprises a microcosm of these contemporary political questions and exemplifies the need for re-consideration of three concepts in particular.

The first is our understanding of state legitimacy. The sources of state stability in the context of a Westphalian system of autonomous and independent nation-states simply do not hold under the political circumstances of today. State legitimacy is sustained by a supportive citizenry in Alberta, because the state provides economic well-being, welfare, and security from violence. But there is more. The provincial state must maintain legitimacy among international investors and consumers, and the pursuit of this international legitimacy may contradict the imperative to maintain legitimacy at home. As such, legitimacy maintenance activities by the province have come to be more about concealing information and risks, and less about providing social welfare.

The second, indelibly related to state legitimacy, is the theoretical notion of citizenship; it too has been challenged by globalized social circumstances and environmental and resource calamities. Alberta is in many ways full of contradictions. In a proto-typical petro-state, the citizen is a non-entity, and there are certainly indications of weakness among the Albertan citizenry. Alberta has one of the lowest voter participation rates in the Western world, and a high proportion of temporary foreign workers who do not even enjoy the right to vote. But, members of this citizenry have nonetheless shown themselves to be active and critical participants in political debates, including the tar sands, and the civic community that they identify with stretches far beyond provincial and national boundaries.

D.J. Davidson and M. Gismondi, *Challenging Legitimacy at the Precipice of Energy Calamity*, DOI 10.1007/978-1-4614-0287-9_7,
© Springer Science+Business Media, LLC 2011

Finally, the ideologies supporting our social institutions, and their means of evolution, must necessarily be at the heart of any analysis of macro-social change.

In all three cases – legitimacy, citizenship, and ideology – these concepts are founded upon sets of boundaries that have come to be entirely fictional – presumed boundaries between state jurisdictions, boundaries between economic development and its environmental consequences, and boundaries between ecosphere and society. In many ways, it was ideology, not economics, which ensured the tar sands' eventual development. A westernized worldview of frontier individualism, a utilitarian view of ecosystems, and confidence in continued progress supported decades of investment in research and marketing by the provincial state, public investments that were crucial to eventually attracting the interest of private capital.

Many aspects of this ideological package have lost their hegemony, however, subject to challenge by a new globalized constituency to which the provincial state is now beholden, exacerbated by the practical challenges to social innovation caused by climate change and peak oil. Furthermore, Albertan citizens themselves are increasingly finding contradictions in this ideological package. Key to this shift has been a series of visual interventions by individuals whose bodies of photographic work presented never before seen industrial vistas of the tar sands. These works broke through the controlled corporate and state images available, and because of the circulating capacity of the Internet have provided a global flow of stills and videos of the scale of industrial operations and its ecological impacts. The images circulate at different scales and among different publics-Don van Hout's (2007) personal canoe journey down the Athabasca River in 2007; Louis Helbig's aerial photography shot from his own airplane as he flew over the "restricted airspace" above tar sands operations; or the disturbing industrial landscape photography of Edward Burtynsky (2011) and his world renown study of The End of Oil with wide scale images of Alberta's tar sands operations prominent in the exhibit. Available extensively, these and many other still and moving images have been repurposed, re-circulated, and sutured into critical oppositional discourses. If an image says more than a thousand words, and we think they do, then the tar sands tale has been told to hundreds of millions of people, who, drawn to them like a lighthouse, may just see a common future and basis for global citizenship.

The Fragility of Legitimacy

Of central importance to the prospects for reflexive social change, whether pertaining to the development of the tar sands or to society's relationship with energy more broadly, are the ways that institutions characterizing that order gain, maintain and defend legitimacy. Legitimacy is technically defined as being in accordance with established or accepted explanations, patterns, standards, and rules. The concept has been employed by students of politics to describe the collective societal approval of the practises of a particular institution, such that the influence of that institution over others is seen as justified. This presumes a relationship of mutual dependence between an institution making certain behavioural concessions for the benefit of a particular audience or audiences, and the audiences that institution depends upon for critical resources (votes, purchasing power, information, etc.). We often view the institution as the power-holder in this relationship, and in many ways this has borne out. But mutual dependency empowers audiences to *refuse* that legitimacy (Suchman 1995) and most researchers tend to emphasize the relative agency of one or the other – institution or audience. Those focusing on institutions highlight the extent to which cultural beliefs shape institutions that become legitimated (e.g., DiMaggio and Powell 1983; Meyer and Rowan 1977). Others focus on the strategic efforts of institutional actors to extract legitimacy from their social environments (e.g., Ferrari 2007; Lazuka 2006; Davidson and MacKendrick 2004; Ashforth and Gibbs 1990).

One can contemplate legitimacy in relation to all manner of authority systems, but states in particular draw our attention here – in this case the Canadian state. Because we are focusing on land use, the state of interest is Canada's regional representative, the Province of Alberta, to which authority for land use decisions has been delegated. State legitimacy is premised upon a relationship between a governing body and a citizenry; one that is more formalized than is the case for other institutions and their audiences. State legitimacy authorizes levels of control over the social order that no other institution enjoys. The legitimacy of a state thus includes a concession of the right to use force, albeit within limits that are not always clear or fixed. The justification for such authority is offered in exchange for social welfare, security, protection of individual liberties, and of course stewardship of natural resources and environmental well-being.

Legitimacy is not solely an explicit social contract between ruler and ruled (Lessnoff 1990) however; it also represents a commonly-held set of beliefs: "a generalized perception or assumption that the actions of an entity are desirable, proper, or appropriate within some social structured system of norms, values, beliefs, and definitions" (Suchman 1995:574; see also Pakulski 1986; Haikio 2007). Legitimacy emanates from the *perceived* moral and ethical validity of the activities over which a state prevails (Schaar 1989), as well as the *perceived* diligence of the state to follow through on its end of the bargain, neither of which is static, or universal. In other words, legitimacy, as with citizenship and ideology, to be discussed herein, exists objectively yet is created (and constantly recreated) subjectively (Suchman 1995), and this means neither legitimacy maintenance nor legitimacy crisis should be considered pre-determined outcomes. While one can accept that, for example, inequities in social welfare constitute justification for a legitimacy crisis as Habermas (1973) argues, a significant level of sociological inquiry has been devoted to evaluations of the perpetuation of objective conditions that could be considered illegitimate according to almost any ethical metric, including not just inequities in social welfare provisioning but also multiple forms of inequality (e.g., Robinson 2009; Rubtsova and Dowd 2004; DiMaggio and Powell 1983) and political violence (e.g., Hollander 2009). Societies have expressed a remarkable propensity to allow horrendous violations of social contracts to persist. This quandary is pertinent to our research into the public acceptance of ecological and ethical contradictions of tar sands' expansion, toward which some concerned citizens and scientists have expressed what Amsler (2009:116) describes as "a sense of incredulity that people would not act to avert the possibility of unprecedented environmental and social catastrophe."

Researchers often attribute the public's non-response to a calamity to powerlessness, or self-interest. But the absence of mobilized opposition to breaches of the social contract may represent something else entirely – the power of persuasion. Skilful discursive legitimation practises, visual and rhetorical, are capable of nothing short of re-creating reality, to the extent that such practises have been described as a form of magicianship (Freudenburg and Alario 2007; Alario and Freudenburg 2006), whereby what matters most to legitimacy is not what supportive conditions are *present*, but rather that evidence of contradictions is notably *absent*. The visual mis-direction observed by Freidel (2008:238) in the staging of "native" images in oil sands corporate media depictions of First Nation Elders apparently in partnership with the companies, offers an example of what Freudenburg and Alario call discursive distraction or mis-direction practised by the elite in a manner that allows power and differential distributions of privilege to become "less observable" and thereby perpetuated (2007:154).

Such perceptual flexibility is accommodated in part by the subjective nature of legitimacy itself. Suchman (1995) describes three distinct components of legitimacy – pragmatic, moral, and cognitive – all of which are subject to discursive manipulation, but to varying degrees. The first depends on the ability of an institution to meet the instrumental expectations of its constituency, such as unemployment insurance and law and order. Moral legitimacy refers to normative evaluations of the appropriateness of an institution's practises, according to the cultural context an institution functions within. Cognitive legitimacy refers to deep-seated societal acceptance of the necessity or inevitability of particular institutions and its practises. When an institution or its postulates become *taken for granted*, it enjoys cognitive legitimacy, and by virtue of its very given-ness this is the most difficult form of legitimacy to unseat (Johnson et al. 2006). One particularly useful means of maintaining legitimacy in this case would be to distract attention away from clear breaches in instrumental legitimacy, an increase in cancer rates downstream from the tar sands, let us say, by drawing attention to the moral necessity of tar sands development in order to curb terrorism, or provide jobs that feed families.

In many ways, Alberta can be seen as a state carrying a big stick, but nonetheless must tread carefully on the discursive terrain of global energy politics. The globalization of markets and polities contributes to the evolution of the social contract. To whom states are accountable has shifted, creating additional avenues where legitimacy crises may unexpectedly erupt (Nye and Myers 2002). Each of these audiences "influences the flow of resources crucial to the [state's] establishment, growth, and survival, either through direct control or by the communication of good will" (Hybels 1995:243), rendering the state vulnerable to rapid, unanticipated changes in the mix of constituent demands (Suchman 1995). To a certain extent, this complex global terrain is not new to states like Canada (and Alberta). Exporters of raw materials have always been sensitive to their legitimacy in the international marketplace.

Because staples commodities are virtually interchangeable – the performance of gasoline derived from Alberta tar is indistinguishable from that derived from Venezuelan oil to the average consumer – producing countries face a perpetual buyer's market. Exporters are thus vulnerable to reputational threats that may lead to consumer boycotts and investor hesitancy. Moreover, Canada's geographic expansiveness renders regional governments that control energy resources and profits (like Alberta) open to criticism from citizens in have-not regions.

The legitimacy of staples states is monumentally more fragile when, as is increasingly the case given the scale of contemporary extractive enterprises, removal and processing activities have implications for global environmental processes. Natural resource development has always posed a legitimacy quandary for states. Such activities require explicit state facilitation, yet states are liable for subsequent disruption to air, land, and water systems (Bunker and Ciccantell 2005), disruptions which tend to be inequitably distributed (Mohai and Saha 2006). Growing awareness and concern in global public spheres of the environmental consequences of such actions may enhance the potential for legitimacy crisis for that same state. The persistence of such state activities in this context demands concerted efforts on the part of state actors to maintain legitimacy. As discussed in Chapter 2, the obscurity of environmental degradation renders environmental politics particularly vulnerable to discursive manipulation, even in comparison to other political realms (Freudenburg et al. 2008; Finlayson 1994; Hajer and Versteeg 2005). And, we have argued that, because authority for licensing is formalized by jurisdictional boundaries, this places even a remote or frontier provincial state, like Alberta, in a position of disproportionate power, *especially if* that state offers what is perceived as an essential and declining resource, like oil.

State engagement in tar sands development requires two forms of critical resources: the support of voting citizens, and financial investments. The list of elements with the capacity to *threaten* either of these critical resources for staples states in a complex global system is a dynamic and ephemeral category, defined by multiple information sources both scientific and otherwise – a complex system of communication networks that convey them, the unanticipated emergence of global competitors, and so on. In global economic sectors like oil, capital can be quite fickle to the propensities of global social (and consumer) sentiment, new scientific and technological information, and volatile commodity prices. As well, given global communication networks, state actors do not have as much control over their message as they might like – which serves as a significant weakness in the ability to defend legitimacy. Legislative Assembly debates, Premier's speeches, and graphic images of the destruction and waste caused by tar sands operations are readily uploaded onto digital communication networks for purview by multiple other audiences. By virtue of their ability to also engage in direct discourse with investors, Albertans, and each other, global audiences have the capacity to threaten state legitimacy.

Evidence of legitimacy maintenance activities by state actors in Alberta's early days was transparent and directed primarily toward attracting investors. Local support was seemingly taken for granted, and the scale of environmental consequences, even the scale of operations that we see today, was at that time inconceivable.

Early efforts by Alberta brought national recognition to the industry when the country began to realize its dependence on foreign oil; and in the 1970s the Canadian postage stamp image of the tar sands giant bucket wheel engraved the industry into the Canadian resource psyche. By the 1980s, public support for the Alberta state received a further boost by the presumed "threat" posed by federal intervention via the proposed National Energy Program. The experience solidified provincial identity, and the ideology and practise of western regionalism energized the tar sands development enterprise (this is *ours* to do with as we wish). Today, legitimacy is under threat from multiple sources: local citizens who have good reason to question whether they are receiving a fair share of the "benefits" on offer, and those who are paying a disproportionate share of the costs – including in particular those living downstream and future generations who will be stuck with the costs of reclamation. Municipal governments fear that many infrastructure and social service costs have been downloaded, not just at ground zero in Fort Mac, but all municipalities where costs for regular maintenance and service provision have skyrocketed in boom times. Each has called into question the behaviour of the provincial state. Agreements to allow temporary foreign workers into Canada have resulted in patterns that have put Canadian labour at risk, as has overseas outsourcing of modular components of the industrial plants. Likewise, trade agreements such as NAFTA also pose a challenge for legitimacy today, as free trade signatories in the 1980s and 1990s willingly divulged a certain degree of autonomy to act in the interests of citizens, including the ability to reserve oil and gas supplies for Canadian consumers, as well as the ability to enact environmental management measures that could be interpreted as protectionism. On the other hand, corporations, as would be anticipated, balk at any attempt by the government to regulate or further tax the oil industry (Fig. 7.1).

To justify and legitimate states' actions, state agents invoke two key privileged accounts: that tar sands development is necessary for Canada's economic well-being and that Alberta oil supplies sustain global political security. But the underlying assumptions for both of these, which proponents have attempted to sweep under the rug, are questionable. Evidence of both the privileged accounts and concealed assumptions have been discussed in other chapters, but they bear summarizing here. In brief, on the economy, here is the information that tends to get swept under the rug.

- The jobs provided by the tar sands are mostly temporary, and a large proportion of these are taken by temporary foreign workers, who send their paycheques back home; some skilled construction jobs are also moving overseas, and refining jobs to the United States.
- The tar sands operators enjoy the benefit of the smallest royalties in the industry, 1% until completely solvent.
- Development is enabled by heavy state subsidizations, including, to take just a few obvious examples: road building, water and electricity provision and the construction of the infrastructure required to provide them, and research.
- Future infrastructural and environmental costs are passed on to taxpayers.

Fig. 7.1 Calgarians, living in the finance and management capital of the energy industry, woke up to Greenpeace one sunny morning in 2010 with a message hanging from the Calgary Tower highlighting the cozy relationship between the tar sands industry and the federal and provincial governments. © Greenpeace

- The distortions of a boom economy mean the cost of living rises faster than incomes (which are not evenly distributed anyway), translating into a decline in the quality of life for lower- and middle-income families.

On security, here is the information that proponents avoid.

- Development of the tar sands does nothing for Alberta's, or Canada's energy security, because 75% of the product is exported to the United States. This will not change so long as the NAFTA remains in effect.
- The current one and a half million barrels a day exported to the United States makes up a small proportion of the 20 mbd consumed in that country, thus tar sands development has little impact on the degree of North American dependence on oil from the so-called terrorist-supporting countries of the Middle East.
- If anything, tar sands development introduces new sources of local risk of acts of terrorism, with a highly dispersed pipeline infrastructure that is virtually impossible

to defend in its entirety, and the introduction of multiple forms of social anomie that characterize booms towns.

- The spectacular environmental catastrophe that would result should any one of the existing tailings ponds fail, places the entire Athabasca River watershed, stretching from northern Alberta to the Arctic, in a perpetual state of insecurity.

While we mentioned the several audiences relevant to the Albertan state, the citizens themselves remain elemental, as they have the power to remove the current ruling party from office. A handful of examples illustrate how local citizens are calling into question the legitimacy of the provincial state during the oil sands consultation hearings:

> Now, Premier Stelmach at a conference, I think it was yesterday downtown, he gave a speech in which his primary message was we're open for business…. Now, this rang me as – I get kind of frustrated and alarmed when I hear things like that…. I just don't understand how Premier Stelmach could say that we can continue development while protecting the environment…. How can you say that without identifying environmental limits? How can you say that without doing the regional environmental assessments, doing the cumulative environmental assessments? You can't. You can't really say that. So that's my biggest concern with this… we're getting one line out of government, which we can do this, we have the capacity and expertise to continue development without harming the environment, but there's nothing backing that up (Oil Sands Consultations, Calgary, April 24, 2007).

> A second thought I've had since then is reading in the paper a few weeks ago, our Premier Ed Stelmach saying that he's not going to touch the brake on oil sands. So I guess my question is what are we doing here? … If the decision has already been made, why are we here? (Oil Sands Consultations, Edmonton, April 3, 2007).

> A study commissioned by the town of Peace River states that as many as 25 companies are already active in the oil sands east of Peace River. These public consultations are coming after the fact, after millions of dollars have already been invested. Why is the government selling leases to oil companies behind closed doors? Why is construction going ahead when projects have not yet been given formal approval? Why is it that the Energy and Utilities Board is not allowed to consider the cumulative impact of all projects and land uses when it holds hearings? (Oil Sands Consultations, Peace River, April 16, 2007).

> I'm here today because I'm angry. I just wanted to say that flat out. I'm angry that my government is allowing expansion and development of the oil sands, and in doing so I feel jeopardizing my future and the future of every other living creature in Alberta and in the world. I'm angry because I feel that as a citizen my concerns have been waved aside. And I'm angry because I think the government has already made up its mind even before starting these consultations (Oil Sands Consultations, Edmonton, April 4, 2007).

Most ENGO representatives were less inclined to challenge directly the legitimacy of the state and expressed a preference for encouraging behavioural changes in less threatening ways. This written submission to the Oil Sands Panel (submitted April 24, 2007) from the Environmental Law Centre is one exception:

> Both Alberta and Canada must recognize and fulfill their ultimate roles as the legal authorities responsible for the legislative powers assigned to them under the Canadian constitution…. Governments cannot and should not be using mechanisms such as CEMA [Cumulative Environmental Management Association] as an excuse or a shield to deflect their constitutional responsibilities.

The hearings themselves, as a huge exercise in symbolic legitimacy maintenance, backfired: the rapid development that continued outside the forum window made a mockery of the process. The state's insistent confidence in environmental amelioration simply did not hold weight against the stark, visual indications of ecological disruption. At least for citizens who were willing to testify at the hearings, those conditions represented clear contradictions to the moral and cognitive claims to legitimacy by the state. The efforts at persuasion by state agents had lost their effectiveness, and their disingenuous actions represented clear breaches of their social contract with the citizenry. But the extent to which the legitimacy breaches recognized by a handful of testifiers will generate a full-scale challenge to state legitimacy is limited, however, without the emergence of a wider consensus.

Re-Articulating Citizenship

What About Environmental Movement Organizations?

We often look to social movement organizations as the source of significant political shifts. The environmental movement has been subject to much scrutiny, some of it quite enthusiastic. The most significant role played by such agents of change is their capacity for introducing new ideas in the public sphere; ideas carefully crafted, however, by means of framing processes (Snow and Benford 1988). Framing processes are intended to pose a challenge to existing dominant frames, and are strategically developed as a means of mobilizing an expanded support network and thus must resonate with sympathizers (Klandermans 1984). This is no easy task, suggesting a need for acute awareness of the cultural and political climate within which organizations function (Benford and Snow 2000).

Certainly environmental movement organizations can and have played an important role in tar sands politics. Can this network of social movement organizations stop the tar sands, and perhaps move beyond this goal to generate a transition away from oil-dependence? Based on our analysis we are sceptical, however we tend to agree with Gamson (1975:28) that "success is an elusive idea". Even if these organizations cannot stop the tar sands, or foster a transition *directly*, they can nonetheless play a critical *indirect* role, by introducing new ideas and a degree of scepticism into the public sphere (Giugni 1998). As such, we can identify clues in the collective action frames employed. There are three tiers of organizational activity in tar sands politics. First are the well-established regional environmental groups that have been critiquing the tar sands for well over a decade, some of which have found themselves in the precarious position of depending on funding from the very corporations they criticize. These organizations, like the Pembina Institute and the Environmental Law Centre have provided some of the most valuable counter-information into local debates, but by and large have adopted "consensus-building" tactics, designed to ensure a seat at the table, rather than turning

Fig. 7.2 In the summer of 2010, Edmonton employees of the cosmetics company Lush participated in an international protest coordinated with the Rainforest Action Network by donning mock oil barrels in protest of tar sands development. Source: http://www.treehugger.com/files/2010/06/ brave-protestors-strip-to-skivvies-to-protest-tar-sands.php. Accessed January 30, 2011. Used with permission Lush Cosmetics

the tables (Pellow 1999). This tactic demands a form of collaborative frame-making that must be sufficiently challenging to resonate with the views of movement sympathizers, and at the same time compatible with the prerogatives of state and industry elites. Regional organizations have adopted a set of frames that align with this collaborative approach. All make clear that their agenda does not include shutting down the tar sands; rather, they tend to call for more science and technological investments, and compliance with existing laws. As a politically conservative staples state, Alberta is in many ways a hostile political arena for an environmental organization, and thus this strategy makes a lot of sense for any organization with the goal of longevity. Organizational leadership also undoubtedly is aware of the reluctance of many Albertans to bring to a halt what is perceived to be the economic mainstay of the province. The calls for change emanating from and the actions employed by these organizations have been tempered to say the least (Fig. 7.2).

We have also seen a newly emergent cohort of international movement organizations, including the giants like Greenpeace, Natural Resources Defense Council, the Polaris Institute, and so on, which have developed specific tar sands campaigns. Newer, more targeted groups have emerged as well like Tarsands Watch, and Oil Sands Truth. With digital communication networks as their primary dissemination vehicle, these organizations have played a notable role in bringing international awareness to the tar sands, and as such they have had a tremendous influence over the framing of non-local discourses. In particular, they have successfully introduced the powerful "Dirty Oil" frame into popular discourse, replete with negative symbolic connotations pertaining not only to the environmental dregs of operations, but also bringing to mind analogies to "dirty money", "blood diamonds", and the like. Well-established international environmental organizations are focused primarily on the tar sands' greenhouse gas emissions and their implication for Canada's compliance with the Kyoto Protocol, while for the emergent targeted groups, like Corporate Ethics International's rethinkalberta.com campaign, the call for action is to stop the tar sands, by promoting divestment via consumer and tourism boycotting and direct pressure on investors. These frames do make some strategic sense – the most plausible means of pressuring Canada to comply with an international accord intended to address a global environmental problem is via international publics, particularly if those publics can induce their own states to join them. A stop to the tar sands would not pose any potential costs onto international consumers, who can get their oil from other sources (at least for awhile), and thus there are certainly no costs for them to embrace such a frame, but then the likelihood of success is slim considering the very large and diverse set of investors and consumers who would need to be convinced in order to completely close the international marketplace to Alberta bitumen.

The success these organizations have had in creating these discursive openings in the international political arena should not be downplayed; the fact that an industrial enterprise on a remote frontier has gained international stature, inspiring direct actions all over the globe, is nothing short of remarkable. But the very selection of certain frames by definition diverts attention away from others, and those others may be quite significant to instigating a post-oil transition to a low carbon economy. Most notably, these activist frames do not resonate with the frames that have been embraced by that population we consider to be of greatest interest – concerned Albertan citizens.

Our third tier is situated back home in Alberta, consisting of a number of small, loosely organized local grassroots groups have begun to spring up like the Keepers of the Athabasca, and the Friends of the Peace River. The positions taken by these organizations have been far more critical than those of the established organizations, but then their support base is also much smaller. Moreover, the frames employed by grassroots organizations are in much closer alignment with unaffiliated citizens, calling for a moratorium on development and posing direct challenges to state legitimacy, to which we turn our attention below.

The Nucleus of Political Change

If the legitimacy of the Athabasca tar sands – much less our dependence on oil – is to be challenged, it will be via the engagement of citizens. Potential for a re-articulation of traditional liberal citizenship rests on identification among Albertans with citizens from elsewhere, and the acknowledgement of the rights of and responsibilities toward ecosystems and other species. To what extent this re-articulation is occurring in contemporary civil societies has been subject to more theoretical contemplation than empirical inquiry, but the current study provides some insight.

Three fundamental features of citizenship today, often overlooked, are critical to the current inquiry. First, citizens are the state's primary source of legitimacy. We have argued that legitimacy has at least as much to do with the success of the privileged at maintaining the dominance of particular ideologies as it does the ability of the state to ensure the provision of constitutional rights in any substantive form (Young 1990). Second, citizenship implies both rights *and* obligations – a criterion easily overlooked when immersed in Western political cultures that place far more emphasis on individual rights than on community well-being (Smith and Pangsapa 2010). This is certainly not a universal expression of citizenship, however. Numerous cultures in other places and times prioritize community responsibility over individual rights. Community responsibility demands consideration of the impacts of one's behaviours on others, which effectively qualifies the determination of individual rights. Whether Albertans privilege their "right" to produce the raw materials located within their jurisdiction, or instead privilege the costs borne by others as a result of that development, defines a primary mechanism through which tar sands development is deemed legitimate or not. This is not an absolute determination, however; it depends highly on just who those Others are. If those Others (First Nations, future generations of Albertans, victims of global warming in other countries, or other species) cannot be clearly identified as members of a shared community of citizens, then consideration for them will lie outside the realms of obligation, requiring expressions of virtue. The nature and degree of lucidity of the impacts in question also matter; certainly in the area of environmental politics, negative impacts can fairly readily be disregarded or explained away because of the limited ability to perceive and assess impacts without specialized training and equipment.

The very definition of citizenship is subjective, undergoing continual re-articulation in the political sphere. Smith and Pangsapa (2010:58) identify this process as *citizenization*: "the way we define and articulate what it means to be a citizen is in process, provisional, and never completed". These authors suggest that we may learn much more about citizenship by focusing on those moments when meanings of entitlement and obligation are produced, contested, and deliberated upon (Smith and Pangsapa 2010; Roche 1992; Christoff 1996; Smith 1998, 1999). Two particularly notable modes of re-articulation are relevant to the current case and will be discussed in greater detail – the purported globalization of citizenship, and the emergence of awareness of environmental well-being as a crucial pre-requisite to human well-being.

A Global Citizenship?

The disintegration of national membership as a key criterion for citizenship has been remarked upon by several scholars. One driving force has been the withdrawal of the state from some citizenship entitlements, which has supported a certain degree of dilution of citizens' loyalty to the state (Sassen 2009). At the same time, growing global material inter-dependence has generated what David Held refers to as "overlapping communities of fate", or what Ulrich Beck calls a "global class of victims", linked together by, for example, a common set of risks associated with the deposition of hazardous wastes, dependence on increasingly scarce raw materials, or vulnerability to global economic turmoil. Most crucially, the growing capacity for international-scale deliberation has generated a public sphere that has become increasingly global (Fraser 2009) and "rhizomal", or highly networked (Castells 2009), creating opportunities for the sharing of such experiences, and the creation of community identities that are not bound to nation-states. The ascription to human rights as a universal principle has been identified as a central bonding mechanism across such globalized communities (Ong 1999; Sassen 2009).

The emergence of globalized identities has challenged the presumed exclusivity of citizenship. According to Hitt (1998:315): "the argument that all human identities are formed oppositionally and that a global human identity is impossible because there would be no 'other' is neither logically defensible nor empirically verifiable". If we consider the capacity to deliberate publicly on contemporary political issues to be a key expression of citizenship (Bohman 2007), then we are without question observing the emergence of new forms of citizenship that represent new forms of legitimacy challenge.

Such observations are certainly encouraging, given that our very ability to address global problems depends on personal identification with a global community (Hitt 1998). Indeed, new research has shown that individuals who perceive themselves as sharing a bond with all of humanity are more likely to exhibit an essential civic virtue – altruism (Munroe 1998). On the other hand, our enthusiasm for an emergent global citizenship with the potential to re-shape politics, including energy politics, needs to be tempered. While identification with a common "imagined" community, connected in large part through media (Anderson 1991) and the capacity to deliberate, is necessary, it is by no means sufficient to constitute citizenship. Moreover, the extensive presence of even these elements is questionable. The prevalence of ethnic violence alone challenges notions of global community, as does the ever-growing gap between rich and poor. The capacity to deliberate as well is far from universal, and while many in civil society sincerely express representation *of* those unable to participate, this is no substitute for direct participation.

Most importantly, to be considered a citizen, an individual must also have the capacity to withhold legitimacy; and while we can observe multiple avenues of influence that members of global society can wield over a particular nation-state, these avenues should by no means be generalized. While non-Albertans can have tremendous influence, via markets in particular, their political influence on individual states other than their own remains indirect. From this perspective, Albertans' formal citizenship status represents a higher degree of potential political power, and given that the state on which they bestow their allegiance prevails over an enterprise with tremendous global-scale impact, these three million individuals carry a particularly heavy burden. Globalization has in effect generated not *a* global citizenship, but multiple citizenships, each with differential forms and levels of influence over those in power.

Ecological Rights and Responsibilities

Environmental or ecological citizenship represents another significant avenue through which our understanding of citizenship is being re-articulated, as citizens everywhere increasingly "exercise their personal rights to environmental justice, while simultaneously extending their duties to the environmental arena" (Wong and Sharp 2009:38, citing MacGregor and Szerszynski 2003). Environmental rights and responsibilities introduce a significant expansion of the concept of citizenship, the rights in question being expanded beyond traditional understandings of equitable distribution of material goods and democratic participation, to include the right to essential ecosystem services, and the assumption of responsibilities including the environmental impacts on others of one's personal behaviours.

This, according to some, is the linchpin of a global citizenship. As noted by Smith and Pangsapa (2010), only in the past decade or so have a multitude of political campaigns been defined as environmental at all, including issues as varying as traffic, food safety, urban renewal, and indigenous peoples' rights. Environment, in short, has begun to permeate multiple political theatres and this provides an opportunity for global coherence. As feminist scholars have been saying for years, the personal is indeed political, including the food we eat and the air we breathe. It is the emergence of this political sphere, in fact, that many argue is the key mode through which citizenship has become globalized (Dobson 2003; Smith and Pangsapa 2010). Through growing ecological awareness of the global extent of many commons, including the oceans, atmosphere, water cycle, and the like, we increasingly recognize that one individual's bond to another individual outside her country is generated not solely by an identity with humanity, but also by the ecological ties that bind. Such material binds evoke not just community but citizenship. Ecosystem services challenge our ability to socially construct boundaries around membership in politically prudent but ecologically arbitrary ways. What goes around comes around, as they say, and while a given group may be conveniently located upstream vis-à-vis one valued flow of ecosystem services, that same group will inevitably be downstream in others.

Of course this is not meant to imply equity in any way – some groups are far more "downstream" than others. The Inuit, Métis, and First Nations peoples inhabiting northern Canada are a case in point, being at the depository of global wind and water circulation currents that bring all manner of effluent into their food sources, while the most that they "take" from global Others include members of various Arctic mammals for food. This is by no means a particular instance of disproportionality in the distribution of the benefits and costs of environmental destruction. Despite the resonance of discursive frames depicting all members of society equal in the benefits and risks of environmental destruction, as William Freudenburg (2005) has shown, nothing can be further from the truth. To the contrary, if one looks, as Freudenburg does, at the emission of air pollutants from the chemicals and minerals sector in the United States, one would be hard-pressed to imagine a more *in*-equitable distribution. When evaluated in terms of the gini co-efficient, whereby a value of zero would represent perfectly equitable distribution, and a value of one representing complete inequity, the distribution of responsibility for emitting toxic air pollutants among mining and chemicals companies is characterized by a gini co-efficient of 0.975.

In the case of the tar sands, two of the most prominent flows one can observe in this system that characterize the disproportionality of this enterprise include the flows of benefit and of risk. The former carves an unambiguous path southward, both in the form of corporate capital gains (which also increasingly head east to the Asian and European headquarters of energy corporations) and in the form of the commodity itself, which enriches the energy affluence of United States citizens. The most noxious form of risk, conveniently for the beneficiaries, flows in an equally unambiguous northerly direction in the form of toxic waste, following the path of the Athabasca River.

Aboriginal peoples living downstream from tar sands operations have become increasingly alarmed, and mobilized, in response to deteriorating health conditions experienced in recent years. In the National geographic Photo Essay on the Oil Sands is Emma Michael, whose story resonates with many Northern residents. Breast cancer survivor, she stands graveside in the cemetery in Fort Chipewyan, Canada. Her brother died of colon cancer 2 years ago, her sister died of a bone-blood cancer 6 months ago and her mother died of cancer 4 months ago. See http://ngm.nationalgeographic.com/2009/03/canadian-oil-sands/essick-photography.

Such inequities certainly raise the spectre of environmental injustice, which, according to many political theorists is a defining feature of an environmental citizenship. Many sociologists have expressed a significant level of enthusiasm for the political efficacy of environmental justice movements to create inroads in areas where traditional environmental organizations have been barred (e.g., Capek 1993; Bullard 1993). The linkage of environmental concern to equity and health issues certainly breaks significant ground not previously breached by traditional environmental organizations. Would an environmental justice mobilization have the potential to unseat the tar sands though, and direct a global transition away from Murphy and Hall's energy cliff? To date, environmental justice movements have largely emerged in response to industrial irresponsibility, causing harm to groups of citizens who are often marginalized already. An environmental justice frame thus far has *not* motivated a sense of responsibility and instead tends to be focused on the rights of victims in relation to particular localized infractions. As a result, they are generally effectively countered by incremental environmental management – perhaps in rare instances motivating national legislation on hazardous waste, but far more often leading to isolated judicial actions against individual facilities.

Citizenship and the Tar Sands

A brief overview of Alberta citizenry highlights the extent to which citizenship is not only undergoing continuous re-articulation, but also is markedly *non*-universal. There are multiple forms of citizenship among the residents of Alberta, each with differing forms of political influence. This includes, on one end of the spectrum, residents living near projects existing and proposed, who according to provincial legislation have special "directly affected" status when it comes to expressing concerns about those projects. Then there are the silent ones, those non-local workers who either by virtue of their tenuous employment, foreign status, fear of reprisal, or simple lack of personal investment in a place not called home, are notably disengaged in public discourse on the tar sands. Including those *indirectly* employed in the energy industry, this group includes some one in six residents of Alberta. In many ways these conditions are not unique to tar sands development, but rather are endemic to contemporary staples industries. The pace and scale of development in modern staples economies means that producing areas will never have a sufficient resident skilled workforce, and would not have the time to train one. So, primary industries become dependent on a mobile workforce, and that mobile workforce becomes dependent on perpetual relocation, ever following that next boom.

Then we have Aboriginal peoples, who themselves have differing levels of status, including registered First Nations who enjoy a formalized relationship with the federal government defined by treaties that include the right to subsistence lifestyles, and can make claims against the federal government when those treaty rights have been infringed upon. The Métis, descendants of marriages between Aboriginals and European settlers, were marginalized during the treaty-making era, but since that time have secured recognition, particularly from provincial governments. In addition, there remain numerous non-registered Aboriginal groups who do not have any formally recognized rights and continue to fight for claims to their traditional territories. The rights enjoyed by First Nations and Métis include the right to "consultation" regarding any proposed activity taking place on traditional territories, giving them what at least on paper constitutes influence above and beyond that enjoyed by non-Aboriginals. Unlike most European descendants, the cultural beliefs of most Aboriginal groups espouse responsibility for future generations, and this responsibility often motivates political engagement. For some, allegiances to both the federal or provincial state, and the non-Aboriginal citizenship, are tenuous given that their ancestry pre-dates the formation of these political bodies. In all three cases, Aboriginal peoples are particularly vulnerable to the impacts of tar sands development, both because they make up a high proportion of downstream and directly affected residents, and because they rely to a much higher degree than the rest of us on the services provided by their local watershed for food and livelihood provision. And all are understandably tired of the efforts needed to counter the repeated infringements on their rights, whether formally recognized in a treaty or not, by state-endorsed industrial development.

Those critical voices that do emerge are worth listening to carefully. As noted earlier, Alberta citizens are in a key position to influence the fate of this local enterprise with global reach – they are the ones with the formal, constitutionally ordained power to withdraw the legitimacy of the provincial state. While global Others can apply pressure, they cannot vote the current leadership out of office, and those testifying at the oil sands consultations, supporting grassroots protest organizations, writing letters to the editor, and so on, represent that body of reflexive individuals in Alberta willing to support change, and can play an important leadership role in the province.

We can observe a distinct discourse being expressed by unaffiliated citizens in the hearings, one that has expressed responsibility for a global citizenry, and has held the state accountable for their impacts on these global Others. This discourse, despite the lack of coordination appears remarkably congruent. The strong themes among citizens testifying at the Oil Sands Consultations, which have been evidenced throughout earlier chapters, include the following:

1. Moratorium on new projects until sufficient planning and safeguards are in place, including attention to cumulative environmental impacts.
2. This is our resource, and we are not getting our fair share; the benefits accrue to the few and the costs to the many.
3. The rapid pace of development is out of control; we do not buy the belief that the economy will automatically "fix" things: we need planning.
4. There is no validity to claims of no negative environmental impacts; those impacts are not even being adequately assessed.
5. The state's interest in hearing citizens' concerns is not genuine, as evidenced by continued rapid development.
6. Our global reputation is being tarnished.
7. We have a responsibility for the global impacts of our actions.

Here are just a handful of representative examples of citizens engaged in a re-articulation of citizenship:

> I speak as a citizen of Calgary and of Alberta, but also as a citizen of Canada and of the Earth which is, after all, our only home. ... Ultimately, the issue that we face in considering the future of the oil sands in Athabasca is not economic or political, or even social. It is a moral and ethical question that strikes right at the heart of who we are (Oil Sands Consultations, Calgary, April 24, 2007).

> Having visited communities in Alberta in the heart of the Tar/OilSands, and the sites themselves, I am embarrassed and disappointed that the province, and our country is allowing such unchecked booms and busts, such short-sighted long-term devastation to go on. It is time for leaders to wake up and take notice that the world is watching, and Alberta development policies are behaving like a drunken teenager, disrespectful of his family, which in this case are Canada and the globe (Oil Sands Consultations, Written submission, April 23, 2007)

> The public, the private sector and government must consider the ethical and moral responsibility for protecting the environment as well as promoting the economic and social development of the Alberta oil sands (Oil Sands Consultations, Fort McMurray, September 27, 2006).

Again, although ENGO representatives did not tap into this citizenship frame to a great extent, there were some notable contributions:

> I think we've got a responsibility to our own people and to the rest of the world frankly given the privileges that we enjoy to show some leadership in dealing with what is probably the greatest threat facing human kind of the 21st Century: the global climate crisis (Oil Sands Consultations, Fort McMurray, September 28, 2006).

The Implications

How is (or is not) this set of concerns entering into political discourse? We are not in a position to argue that the small proportion of Alberta's citizens who participated in the Oil Sands Consultations and other public reviews will by virtue of their contributions change the course of tar sands development. But the message conveyed by these individuals does indicate the presence of a distinctive resistance frame with resonance on several levels – in terms of rights, responsibilities, virtues, and ecological consciousness. And it is being expressed by Albertans themselves, who have the unique power to challenge directly the endorsing state apparatus. This discourse is meaningful in other respects as well, offering suggestive evidence that the tar sands is serving as an entry point into the re-articulation of citizenship, and other similar enterprises of global catastrophic merit may well do the same. What these narratives suggest is that, although the state has pursued industrial development ostensibly for the purpose of benefiting Albertans, the contradictions embedded in this privileged account have become sufficiently blatant that many citizens have come to view the state's actions as infractions of the social contract.

As well, because the actions endorsed by the Alberta provincial state are impacting a community that is much broader than its formally defined citizenry, it must assume culpability for its infringements on the rights of the members of that broader community. Hence, a de facto expansion of the "citizenry" to which the state is being held accountable, morally and ethically if not formally, has been remarked upon by citizens themselves. This has invoked action in the pursuit of state accountability, not solely in response to violations of personal rights – although many have indeed remarked upon the personal consequences of development – but even more so on behalf of those Others. Awareness of global-scale environmental and ecological destruction has in this case served as the means by which these citizens have come to perceive the global stretch of the communities with which they are related. But there is another element at work here that appears to motivate action, that might be seen as less virtuous as altruism but nonetheless an effective motivator. In an increasingly globalized, networked society, that community from which an individual receives signals of acceptance, identity and sanction has expanded accordingly. As such, Albertans have expressed surprise and dismay over their guilt by *association* with the tar sands within that global community.

Tar Sands as Ideology

Without a shift in the ideologies supporting continued growth in fossil fuel consumption, it is questionable whether even a substantial political mobilization will support outcomes that can change the current disproportionality equation. As elusive as it is to measure directly, macro-social change is premised on shifts in ideology, because it is essentially the ideologies themselves, rather than the privileged interests who benefit from them, which must be accorded legitimacy in order for social institutions to persist (Habermas 1975). According to Fairclough, "ideologies are representations of aspects of the world which can be shown to contribute to establishing, maintaining, and changing social relations of power, domination, and exploitation" (Fairclough et al. 2003:9). Privileged actors are in essence accorded legitimacy (or not) on the basis of the ideological paradigms they embrace, and once achieving privileged status, they are then in a position to engage in legitimation activities that serve to build further support for those paradigms.

Ideologies are the quintessential social construction; however, they are by no means constructed in a vacuum. Ideology reflects an ongoing interplay between circumstances and events and our interpretations of them. Sociologists have a tendency to emphasize their stability, and for good reason: they are socially reproduced in everyday norms and practises. An ideological paradigm is particularly stable when it achieves hegemonic status, such that it comes to be unquestionably embraced, not only by that group that benefits directly but also by those who pay the costs of support. To describe an ideology as hegemonic, that ideology must have achieved the status of the taken for granted, assuming a seeming ontological status. The fact that such ideologies serve the interests of *particular* groups at the cost of others is concealed behind the guise of generalization. We have discussed such phenomena throughout the book, in reference to cognitive legitimacy, for example, and the "unquestioned assumptions" that under-gird those privileged accounts that support privileged access to resources and waste sinks (Freudenburg 2005, 2006).

But even hegemonic ideologies are vulnerable to challenge, when the contradictions embedded in those ideologies become revealed (Gramsci 1971). Scholars tend to focus on external counter-hegemonic forces (e.g., Evans 2000; Chin and Mittelman 1997), but as O'Connor (1973), Arrighi (2005) and others have noted, the very pursuit of some enterprises with the blessing of hegemony can lead to their own undoing. As noted by Kuhn (1962), however, regardless of the evidence against it, a given paradigm can persist if no alternative paradigm is available to replace its position of dominance. In a sense, the stability of an ideology can be seen as representing a spectrum from hegemony on the one hand, to a state of open debate over multiple competing paradigms on the other (Cox 1983), with prospects for social change increasing as we move toward the latter end of the spectrum.

Conceptualizing Processes of Ideological Change

Conceptually, ideological systems change when: contradictions in dominant ideologies are revealed, social agents respond to those revelations, and new paradigms are available to replace the old ones. In reality, predicting the more specific conditions that allow for such shifts is more difficult. Ideologies are continuously evolving, albeit in some cases along very slow, unanticipated, and seemingly circular pathways. Ideologies, as with capital, fuel, labour, and information, are subject to the system of flows defining global society, and thereby may evolve more rapidly and unexpectedly today than has been the case throughout history. This is a two-way

Fig. 7.3 Mallard drake ducks floating dead on their sides in a black bitumen mat on the surface the Syncrude Aurora tailings pond. The images were taken by Mr. Todd Powell of Alberta Fish and Wildlife, a department in Alberta's Ministry of Sustainable Resources. Over 1,600 ducks were eventually found dead in 2008 on the tailings pond. Mr. Powell's testimony and his images were key evidence in the court case that saw in which Syncrude was found guilty in 2010. Syncrude, a multi-billion dollar company, was fined three million dollars. Powell's testimony and this image appear online at "Over 200 Ducks Dead in Toxic Syncrude Tailings Pond", *Edmonton Journal*. 26 October 2010

street of course – such flows have enabled the rapid colonization of the world's economies by trans-national capital (Gill and Law 1989), but those same networks can also serve the interests of counter-hegemonic forces (Evans 2000).

From this systems perspective, we can postulate a number of more specific sets of conditions that *may* culminate into ideological change. First, globalized communication networks enable citizens to "put the pieces together", drawing connections among independent events more readily, as news of localized events is widely and rapidly disseminated. As a result, the conjuncture of multiple independent events that come to be seen as inter-related can render those isolated events more meaningful in and of themselves, and likewise the culmination of events can awaken observers to the contradictions common to each. Consider in this case the mass duck killing on Syncrude's tailings pond on (in April 2008), which was followed by a poorly-advised defence on the part of the guilty corporation that brought even more fanfare to the event. The 1,606 ducks that perished would not likely have raised an eye under many other circumstances, but they raised international ire because they met their fate on a tailings pond, which had already achieved graphic notoriety on the world-wide web (Fig. 7.3).

BP's oil rig in the gulf blew around the same time as the Syncrude case went to court, along with the failure of an aluminium plant tailings pond in Hungary, all

Fig. 7.4 "Alberta's $25 million PR campaign lands in toxic pond" by Malcolm Mayes. 2008-05-02, with permission Artizans.com

against the backdrop of Hollywood blockbuster film *Avatar*, which was likened so effectively to the tar sands narrative that director James Cameron was compelled to pay a personal visit in Fall 2010. Such coincident events might just serve as a cumulative "aha" moment for Western consumers, creating opportunity windows for ideological change. Then the ducks themselves sacrificed another 500 or so of their kin to the cause, landing on another Syncrude tailings pond not 3 days after the company's court settlement for $3 million dollars in late 2010 (Fig. 7.4).

Such opportunity windows are far more likely to instigate change if they are capitalized upon by social movement organizations, however (McAdam 1982). In this case, there were some organizations that, by coincidence or design, mounted some very complementary campaigns, including, for example, an organizational endorsement for *Avatar* for an Oscar signed by Sierra Club, Suzuki Foundation, and many others; and a photo exhibition featuring the tar sands in London in late September 2010 that received significant media coverage, sponsored by Greenpeace. Greenpeace

Fig. 7.5 A full page advertisement entitled Canada's Avatar Sands appeared in *Variety Magazine*, March 5, 2010. Source: http://cdn.fd.uproxx.com/wp-content/uploads/2010/03/Avatar_Sands_ Variety_ad.jpg. The copy indicated that indigenous people were threatened by the oil sands industry, likened Shell and BP Sands Banner at Calgary the Sky People in Cameron's movie, and presented an image of a huge "helltruck" bent on removing the unobtanium from the tar sands. The ad encouraged Cameron to visit Fort McMurray. Image accessed December 18, 2010. On the James Cameron visit to the region see http://dirtyoilsands.org/dirtyspots/category/avatarsands/

also launched a mock travelingalberta.com site touting the tourism potential of Alberta's tailings ponds, among other physical features of tar sands development. The majority of these campaigns were directed toward international sympathizers; social movement activity was much quieter closer to home in Alberta (Fig. 7.5).

Other facilitating factors include the mobilization of dissent "from the middle" rather than the committed social activists from whom we have come to expect dissent.

In this instance, unaffiliated Alberta citizens across the age and occupational spectra, municipal elected officials, even former leaders of staples development, have emerged to express their scepticism of the current state-endorsed development vision. Most recently, the voices of concern emanating from the academy have taken centre stage; the publishing of a rather scathing scientific documentation of the impacts to water quality in the *Proceedings of the National Academy of Sciences* has lent credence to earlier papers printed in less prestigious scientific journals, and the growing scientific acceptance of peak oil expressed by members of the academy have forced fossil fuel proponents, who until recently could effectively disregard such warnings as the ranting of alarmists, into defensive positioning. And although individual workers have not been outspoken critics organized labour has been surprisingly scathing in their critiques of the growth paradigm dictating tar sands development.

The degree of complementarity among multiple ideologies invoked in relation to particular projects can also serve as either a source of stability or of challenge. In the current case, the three ideological paradigms described herein – Western individualism, neo-liberalism, and human exemptionalism – are remarkably complementary, so much so that challenging one implicitly poses a challenge to the other two. As with all such inter-related phenomena, weaknesses in one may be cloaked by strengths in the other two, or on the other side of the coin, significant breaches in the legitimacy of one may ripple throughout the triad. The paths describing such processes are exceedingly difficult to forecast within a complex network society.

The Silver Lining Specialists

Tar sands advocates were quick to capitalize on the BP oil rig blow-out in the Gulf of Mexico, viewing the tragedy as an illustration of just how green tar sands development is by comparison, and celebrating the likely increased production pressure on Athabasca in response to the anticipated slow-down of production in the gulf. An editorial posted on May 29, 2010 in the *Sun Times* in Calgary, claims that "the most obvious [conclusion to be drawn from the spill] is that the Alberta tar sands have become much more valuable". And this, 10 days earlier, from an editorial posted in the *High River Times*, serving a largely rural readership in central Alberta:

> All that oil floating on the Gulf of Mexico reveals the lie about Alberta's "dirty oil" for what it is. While we get no pleasure from seeing our neighbours to the south contend with an unfolding ecological disaster, it at least puts the relative merits of different energy sources in perspective. No matter what the source, extracting oil from the Earth's bosom is an endeavour fraught with risk.... Environmental extremist groups have mounted an unceasing effort to portray oilsands production as "dirty oil." But their main basis for that claim, that oilsands production emits more greenhouse gases, pales in comparison to the potential catastrophe the United States is facing in Louisiana.

Three Ideological Mainstays

There are several sets of ideologies that have facilitated tar sands development, and are readily observed in discourse. Certainly Western individualism has been invoked in compelling ways, with proponents reaching out to an historic Albertan system of values and virtues that support laissez-faire economics, frontier landscapes and masculine individualism, and a strong populist ascription to political independence. The very extent of the "mega" status of the resource, the multiple tools in use, and the "giga" project status of the industry itself, reflects a particular Western symbolism of endless frontier landscapes waiting to be conquered through the exercise of individual determination and ingenuity. One researcher labelled the tar sands a "megaprogram", on par with to NASA or the Internet (Joseph and Ginton 2010). Culture and media pundits liken it to Chevy Trucks, John Wayne movies and muscle-bound heroes in a manner that invokes a sense of pride and ownership over the enterprise. Such narratives are masterful in that no attempt is made to shield the scale of disruption, it is instead recreated as evidence of our masterful ingenuity and Alberta heritage; we *want* to believe. The very purpose of creating problems is to express our prowess at fixing them:

> To halt growth would be to deny our ability to meet growth's challenges with Alberta's leadership, innovation and entrepreneurial spirit (Stelmach. Speech at the Canadian Energy Research Institute Oil Conference, Calgary, April 23, 2007).

Two additional ideological systems, not quite so Alberta-centric, which because of their very omni-presence throughout the global polity, are especially critical to the prospects for change in energy politics today. They are central to the legitimation of the tar sands, indeed of industrial capitalism in general. One is a relatively recent (re)iteration of a particular economic ideology, neo-liberalism, ascribing to the belief that un-fettered (read, un-fettered by the meddling of governments) markets are rational actors in and of themselves, providing autonomic responses to scarcity, protecting things of value, and most of all enabling all boats to ride the wave of material growth. What makes the market rational, by extension, is the presumed rationality of the behaviours of all of the producers and consumers operating in that marketplace. *We* are the market, and the market is us. And of course there is the pre-requisite that everything in the universe has a (US) dollar value.

As has been shown in Chapter 3, there would appear to be rather clear indications that, as Saul (2005) predicted, neo-liberalism may well be in its twilight. This is most clearly evidenced by the degree of consensus among Albertans that development is "out of control", meaning the market will not simply self-correct, and what is needed is a moratorium on further development. Proponents, not surprisingly, warn of the dire consequences that await:

> I want to make one thing very clear. There has been talk by the federal Liberals – and others – of a moratorium on development of Alberta's oil sands. My government does not believe in interfering in the free marketplace. You cannot just step in and lower the boom on development and growth – in the oil sands or elsewhere. If that were to happen, the economic consequences for Alberta, and for the economy of Canada, would be devastating (Stelmach. Rotary Club of Calgary Valentine's Luncheon Address, Calgary, February 13, 2007).

But such warnings have little merit among citizens who are told of record corporate profits at the same time they struggle to pay for a sky-rocketing cost of living; citizens who have come to agree with MLA Swann that:

> Genuine progress presupposes another principle that the economy fundamentally serves people and the environment, not the other way around (Swann, MLA. Testifying at Oil Sands Consultations, Calgary, April 23, 2007).

One notable source of challenge to neo-liberalism, and increasingly to Human Exemptionalism, discussed further below, is labour. The Canadian Communications, Energy and Paperworkers Union (CEP) have over 150,000 workers in the primary resource extraction economy in Canada, including heavy equipment operators, gas plant and refinery workers, and operations and blue collar jobs in the tar sands. CEP's membership is anticipating how they might adapt to more stringent environmental regulations on the oil sands industry that may impact the pace of tar sands development, and numbers and duration of jobs in Fort McMurray. Framing their concerns in moral as much as economic terms, they have rejected simple zero sum arguments that environmental controls means job losses, and are exploring a new concept of "just transition", to engage in discussions that move society forward:

> Just transition is an approach to public policy that seeks to minimize the impact of environmental policies on workers in affected industries (e.g., through retraining programs, preferential placement opportunities, development, or diversification assistance in affected regions) (Daub 2010:116).

The support and involvement of labour will be crucial to any social transition beyond what the authors have dubbed the prevailing "take advantage of the tar sands while it lasted" mentality, to a more progressive "desire to find ways to even out of the boom and bust cycle and create stable dependable economic futures – a core goal of the CEP energy plan". The CEP's framing of work and climate change as social justice issues includes moral conceptions of environmental sustainability and the distribution and prevention of risks (Daub 2010:133, 116, 124, 125). The concept of just transition may also provide humans a way to imagine the conversion to a new energy age.

Human Exemptionalism

Another enduring ideology embraced by Western societies – one that neo-liberalism is directly dependent upon – is what can be called human exemptionalism – the confidence that Nature can be controlled and put to good (economic) use (Catton and Dunlap 1978). There are numerous frames embedded in this paradigm, such as confidence in technology to reverse environmental impacts, the infinite substitutability of depleted resources, or simply the tendency to presume climate change will have no social impacts. Human history can in many ways be perceived as marked by a progressive relinquishment from environmental/material constraints, with each new breakthrough in disease control or food production offering supportive evidence, and concealing the counter-evidence. As author and tar sands supporter Gordon

Kelly says, "there are many people who have predicted the end of the world, but Mother Earth has been able to soldier on and is still functioning reasonably well" (Kelly 2009:42).

After a century of environmentalism, has this paradigm lost its stature? There are *some* indications of competition among fundamentally differing interpretations of our role in the ecosphere, indeed there always have been, but their resonance among any but a small minority in contemporary society is not so clear, and this would appear to be the case in tar sands politics as well. Industry and state proponents have certainly become shrewd, practised orators, banking on (and enhancing) their audiences' ascription to this paradigm. As simply put by the Canadian Association of Petroleum Producers, "Environmental standards are met regardless of the pace of development", and problems will be solved not by precaution but by teamwork: "I'm confident through the right partnerships between industry, government, and society, we can find sustainable solutions".[1] Even more common are expressions of reliance on technology, as expressed by elected officials and industry alike:

> The challenge of climate change must be looked at in a way that will realistically attack the question. Why not support the search for economical ways to remove carbon dioxide from the atmosphere? Just fix it. There must be a way (Backs, Legislative Assembly, March 12, 2007).

> We think technology can transform our industry into an even greater engine of economic growth while also protecting the environment (Oil Sands Consultations, Calgary, September 28, 2006).

The belief that Nature has no value until usurped by humans is another discursive frame that emerges throughout supporter discourse:

> Over 175 billion barrels in proven reserve in the oil sands and 1.6 billion barrels of conventional oil still exist in the ground. Over 41 trillion – that's trillion – cubic feet of remaining established marketable gas reserves still exist in this province in the ground. Coal reserves are estimated at approximately 34 billion tonnes. Now, that's incredible reserves in this province. But they're all in the ground, and those reserves have absolutely zero value when they're still in the ground (Griffiths, Legislative Assembly, May 7, 2007).

In an ironic twist, the Premier posits that tremendous environmental costs would be borne were we to *stop* development:

> A growing economy ... pays for the infrastructure to support public transit – which in turn reduces personal vehicle use. And it helps fund the research that will provide long-term solutions to the environmental issues we face (Stelmach. Canadian Urban Transit Association Annual Conference, Edmonton, Alberta, May 28, 2008).

> Shutting the door to Alberta oil would mean opening the door to offshore suppliers who don't place as high a premium on the environment (Stelmach. 6th Annual TD Securities Oil Sands Forum, Calgary, Alberta, July 9, 2008).

---

[1] Canadian Association of Petroleum Producers, written Submission to the Oil Sands Consultation Panel, October 2006.

Opponents are not necessarily universally at odds with this paradigm – some testimonials in fact imply subscription to the same set of beliefs, espousing confidence in future technological development, and support for "managed growth", with "social and environmental integrity and economics in balance" (Swann, Liberal MLA, testimony). According to one organizational representative:

> In a "can do" province like Alberta, we could achieve this vision of greenhouse-gas-neutral oil sands by the year 2020.... There are two potential ways in which the oil sands industry could take responsibility for its GHG emissions without stopping development. The first is for the industry to seek technology breakthroughs that significantly cut the GHG intensity of oil sands production. The second is for the industry to offset emissions by purchasing credits that represent genuine GHG emission reductions achieved elsewhere (Oil Sands Consultations, Edmonton, September 26, 2006).

There are certainly some powerful ecological justice frames that have been invoked, particularly in regard to Aboriginal peoples:

> Our community will see no benefits from oil sands developments that correspond to the risks and costs we are being subjected to. Alberta wants us to accept these developments in our backyard, which is the source of our food and water, when Albertans would neither accept such developments next to their subdivisions, nor accept the loss of meat and vegetable sections in their supermarkets. The Woodland Cree are pragmatic people, but we have been placed in a very difficult situation as a result of these developments.... Where does that leave us? How will our people survive after the last of these company's trinkets are gone and we no longer have the land to sustain us? Where will we hunt, fish, and trap for the next 50 years while we wait for the land, wetlands, and forest to regrow? Are we supposed to continue taking trinkets in exchange for ill health and cultural threats? (Oil Sands Consultations, Peace River, April 16, 2007).

> We met with the environment minister, and I remember that meeting really clearly because he sat across from two elders and basically told them that, yes, he can hear what they're saying about the cancer in their communities and what's going on, but he really needed to let them know that the technology up there is flawless, so really what they may be seeing on the ground is, you know, probably not what they really think it is, and that the tailings ponds they're not leaking, none of that. I think that most of us know that there are things going on up there that we can't maybe necessarily explain right now and that people are feeling impacts that they should not be feeling, and as someone who lives in Alberta – or who lives in Canada, which is a first world country where we are a democracy and we have amazing rights and we have a quality of life that most people envy across the world, these are not things that should be happening in our communities, and everybody should have the right to clean air, clean water, and a safe future for their children. They shouldn't fear what they drink or what they catch or what they hunt [ENGO, "speaking on behalf of myself as an Albertan"] (Calgary, April 23, 2007).

Signs of Change?

One could argue that Western individualism is locally hegemonic in that it permeates so much of culture, politics and everyday practises in Alberta (and in some other regions with similar political economies). But its resonance is limited by geography, and in a sense this has become a particular challenge for tar sands proponents, who feel compelled to invoke this paradigm in order to sustain support at home, and yet this paradigm does not resonate with other audiences, such as concerned citizens in Europe, or even in eastern Canada.

Neither of the remaining two ideologies in question here could be said to be ideal-typical hegemonic ideologies, since both are subject to scrutiny. Neo-liberalism has been especially heavily criticized, with alternatives not only proposed but implemented in some countries. The particularly bold pursuit of this paradigm in tar sands development has contributed to the revelation of contradictions thereby creating opportunities for challenge that might not have been available had the proponents not ascribed quite so blindly to the power of the market. The Human Exemptionalism Paradigm, on the other hand, while being subject to scrutiny in at least some circles for over a century, has proven to be quite resilient to that criticism. However, the very nature of non-conventional fuel development requires some rather blatant, grotesquely vivid environmental costs that would take some rather magical discursive reframing to explain away. The provincial state, despite an investment in the tens of millions in portraying their endeavours as green, has not shown itself to be capable of such magical powers. But, as noted by Thomas Kuhn, a paradigm shift cannot occur until a viable alternative paradigm has emerged, and despite decades of dialogue on sustainable development, the precautionary principle, and simple living, all are replete with contradictions of their own. And most importantly, all the alternatives on the table also imply costs, not just to a small privileged elite, but also to material-consuming middle classes everywhere, not to mention a burgeoning global underclass that rightfully demands a significant *increase* in material consumption just to meet basic needs.

While there are certainly many sources of stability in the dominant status of the Human Exemptionalism Paradigm, there are two particular circumstances that are compelling. First, at the regional level, the moratorium narrative is of particular interest not only because support for this narrative has been cross-cutting, but also because it simultaneously challenges all three paradigms described here. Whether proponents of a moratorium have even recognized it themselves, there is simply no consistency in acknowledging that tar sands development poses irreparable damage without concerted cumulative effects management and controls on the pace of development, while also ascribing to any of the three paradigms in question. That this narrative has *not* been capitalized upon by a wider network of movement organizations is unfortunate. Second, the Energy Return on Investment analytical tool offers a completely new means of evaluating society-environment relations that does not rely on collective agreement on the values of altruism or ecological consciousness (which are themselves subject to multiple iterations). EROI analyses provide a shrewd, analytically sound, and scientifically robust means of expressing the costs and benefits of material consumption and environmental disruption in a manner that sheds new light on everything from renewable energy to food production. This, too, could become a significant source of destabilization for the Human Exemptionalism Paradigm, but as with the moratorium narrative, this potential can only be realized if social agents beyond a small group of energy analysts capitalize on it.

References

Alario, M. & Freudenburg, W.R. (2006). High-risk Technology, Legitimacy and Science: The U.S. Search for Energy Policy Consensus. *Journal of Risk Research, 9*(7), 737–53.

Amsler, S. (2009). Embracing the politics of ambiguity: Toward a normative theory of 'sustainability.' *Capitalism, Nature, Socialism, 20*(2), 111–125.

Anderson, B.R. (1991). *Imagined Communities: Reflections on the Origins and Spread of Nationalism*. London: Verso.

Arrighi, G. (2005). Hegemony unraveling—1. *New Left Review, 32*(Mar/Apr), 23–80.

Ashforth, B.E., and Gibbs, B.W. (1990). The double-edge of organizational legitimation. *Organization Science, 1*, 177–194.

Benford, R.D., and Snow, D.A. (2000). Framing processes and social movements: An overview and assessment. *Annual Review of Sociology, 26*, 611–39.

Bohman, J. (2007). *Democracy Across Borders: From Demos to Demoi*. Cambridge, MA: MIT.

Bullard, R.D. (ed) (1993). *Confronting Environmental Racism: Voices From the Grassroots*. Boston: South End.

Bunker, S.G., & Ciccantell. P.S. (2005). *Globalization and the Race for Resources*. Baltimore: Johns Hopkins University.

Burtynsky, E. (2011) OIL. Gottingen:Steidl.

Capek, S.M. (1993). The 'environmental justice' frame: A conceptual discussion and an application. *Social Problems, 40*(1), 5–24.

Castells, M. (2009 [1996]). *The Rise of the Network Society*. Hoboken, NJ: Wiley.

Chin, C.B.N., & Mittelman J.H. (1997). Conceptualising resistance to globalization. *New Political Economy, 2*(1), 25–37.

Catton, W.R., Jr., & Dunlap, R.E. (1978). Environmental sociology: A New Paradigm. *The American Sociologist, 13*(Feb), 41–9.

Christoff, P. (1996). Ecological citizens and ecologically guided democracy. In B. Doherty & M. de Geus (Eds.), *Democracy and Green Political Thought: Sustainability, Rights and Citizenship*. London: Routledge.

Cox, R.W. (1983). Gramsci, hegemony, and international relations: an essay in method. In R.W. Cox and T. Sinclair. (eds). *Approaches to World Order*. Cambridge: Cambridge University.

Davidson, D.J., & MacKendrick, N.A. (2004). All dressed up and nowhere to go: the discourse of ecological modernization in Alberta, Canada. *Canadian Review of Sociology and Anthropology, 41*(1), 47–65.

DiMaggio, P.J., & Powell, W.W. (1983). The Iron Cage revisited: Institutional isomorphism and collective rationality in organizational fields. *American Sociological Review, 48*(2), 147–160.

Dobson, A. (2003). *Citizenship and the Environment*. Oxford: Oxford University.

Daub, S. (2010). Negotiating sustainability: climate change framing in the Communications, Energy and Paperworkers Union. *Symbolic Interaction, 33*(1), 115–140.

Evans, P. (2000). Fighting marginalization with transnational networks: counter-hegemonic globalization. *Contemporary Sociology, 29*(1), 230–241.

Fairclough, N., Graham, P., Lemke, J., & Wodak, R. (2003). Introduction. *Critical Discourse Studies, 1*(1), 1–7.

Ferrari, F. (2007). Metaphor at work in the analysis of political discourse: investigating a "preventive war" persuasion strategy. *Discourse and Society, 18*(5), 603–25.

Finlayson, A. (1994). *Fishing for Truth: A Sociological analysis of northern cod stock assessments from 1977 to 1990*. St John's, NL: Institute of Social and Economic Research, Memorial University.

Fraser, N. (2009). *Scales of Justice: Reimagining Political Space in a Globalizing World*. New York: Columbia University.

Freidel, T. (2008). (Not so) crude text and images: staging NATIVE in 'big oil' advertising. *Visual Studies, 23*(3), 238–254.

Freudenburg, W.R. (2005). Privileged access, privileged accounts: toward a socially structured theory of resources and discourses. *Social Forces 84*(1), 89–114.

Freudenburg, W.R. (2006). Environmental degradation, disproportionality, and the double diversion: reaching out, reaching ahead, and reaching beyond. *Rural Sociology 71*(1), 3–32.

Freudenburg, W.R., & Alario, M. (2007). Weapons of mass distraction: magicianship, misdirection, and the dark side of legitimation. *Sociological Forum 22*(2), 146–173.

Freudenburg, W.R., Gramling, R. & Davidson, D.J. (2008) Scientific Uncertainty Argumentation Methods (SCAMs): Science and the politics of doubt. *Sociological Inquiry 78*(1), 2–38.

Gamson, W. (1975). *The Strategy of Social Protest*. Homewood, IL: Dorsey.

Gill, S.R. and Law, D. (1989). Global hegemony and the structural power of capital. *International Studies Quarterly, 33*, 475–499.

Giugni, M.G. (1998). Was it worth the effort? The outcomes and consequences of social movements. *Annual Review of Sociology, 98*, 371–93

Gramsci. A. (1971). *Selections from the Prison Notebooks of Antonio Gramsci*. Quintin, H. & Smith, J.N. (Eds. and trans.). New York: International Publishers.

Hajer, M. & Versteeg, W. (2005). A decade of discourse analysis of environmental politics: Achievements, challenges, perspectives. *Journal of Environmental Policy & Planning, 7*(3), 175–84.

Habermas, J. (1973). *Theory and Practice*. Boston: Beacon Press.

Habermas, J. (1975). *Legitimation Crisis*. Ypsilanti, MI: Beacon.

Haikio, L. (2007). Expertise, representation and the common good: grounds for legitimacy in the urban governance network. *Urban Studies*, 44(11), 2147–2162.

Hitt, W. (1998). *The Global Citizen*. Columbus OH: Battelle.

Hollander, P. (2009). Contemporary political violence and its legitimation. Society, 46(3), 267–274.

Hybels, R.C. (1995). On legitimacy, legitimation, and organizations: a critical review and integrative theoretical model. *Academy of Management Journal, Special Issue* (Best Papers Proceedings 1995), 241–245.

Johnson, C., Dowd, T.J., & Ridgeway, C.L. (2006). Legitimacy as a social process. *Annual Review of Sociology, 32*, 53–78.

Joseph & Ginton (2010). The Tar Sands of Alberta: Exploring the Gigaproject Concept. Paper presented at Unwrap the Research Conference, Fort McMurray. Oct 22–24.

Kelly, G. (2009). *The Oil Sands: Canada's Path to Clean Energy?* Cochrane, AB: Kingsley.

Klandermans, B. (1984). Mobilization and participation: social-psychological expansions of resource mobilization theory. *Annual Sociological Review, 49*, 583–600.

Kuhn, T.S. (1962). *The Structure of Scientific Revolutions*. Chicago: University of Chicago.

Lazuka, A. (2006). Communicative intention in George W. Bush's presidential speeches and statements from 11 September 2001 to 11 September 2003. *Discourse and Society 17*(3), 299–330.

Lessnoff, M. (1990). *Social Contract Theory*. New York: New York University.

MacGregor, S. & Szerszynski, B. (2003). Environmental citizenship and the administration of life. Paper presented at the Citizen and the Environment Workshop, Newcastle University, September 4–6.

McAdam, D. (1982). *Political Process and the Development of Black Insurgency*. Chicago: University of Chicago.

Meyer, J.W., & Rowan, B. (1977). Institutionalized organizations: formal structure as myth and ceremony. *American Journal of Sociology, 83*(2), 340–363.

Mohai, P. & Saha, R. (2006). Reassessing racial and socioeconomic disparities in environmental justice research. *Demography, 43*(2), 383–99.

Munroe, K. (1998). *The Heart of Altruism: Perceptions of a Common Humanity*. Princeton, NJ: Princeton University.

Nye, J. and Myers, J.J. (2002). *The Paradox of American Power: Why the World's Only Superpower can't go it Alone*. New York: Oxford University.

O'Connor, J. (1973). *The Fiscal Crisis of the State*. New York: St. Martin's.

Ong, A. (1999). *Flexible Citizenship*. Durham, NC: Duke University.

Pakulski, J. (1986). Legitimacy and mass compliance: reflections on Max Weber and Soviet-type societies. *British Journal of Political Science, 16*, 35–56.

Pellow, D.N. (1999). Framing emergent environmental movement tactics: mobilizing consensus, demobilizing conflict. *Sociological Forum, 14*(4), 659–683.

Robinson, J.W. (2009). American poverty cause beliefs and structured inequality legitimation. *Sociological Spectrum, 29*(4), 489–518.

Roche, M. (1992). *Rethinking Citizenship: Welfare, Ideology and Change in Modern Society.* Cambridge, UK: Polity.

Rubtsova, A. & T.J. Dowd. (2004). Cultural capital as a multi-level concept: the case of an advertising agency. *Research in the Sociology of Organizations, 22,* 117–146.

Sassen, S. (2009). Incompleteness and the possibility of making: towards denationalized citizenship? *Cultural Dynamics, 21*(3), 227–254.

Saul, J.R. (2005). *The Collapse of Globalism: And the Reinvention of the World.* Toronto: Viking.

Schaar, J.H. (1989). *Legitimacy in the Modern State.* New Brunswick, NJ : Transaction Publishers.

Smith, M.J. (1998). *Ecologism: Towards Ecological Citizenship.* Buckingham, UK: Open University.

Smith, M.J. (1999). Thinking through ecological citizenship. In Smith, M.J. (Ed.) *Thinking Through the Environment.* London: Routledge.

Smith, M.J., & Pangsapa, P. (2010). *Environment and Citizenship: Integrating Justice, Responsibility and Civic Engagement.* London: Zed Books.

Snow, D.A., & R.D. Benford. (1988). Ideology, frame resonance, and participant mobilization. *International Social Movement Research, 1,* 197–218.

Suchman, M.C. (1995). Managing legitimacy: Strategic and institutional approaches. *Academy of Management Review, 20,* 571–610.

Van Hout, D. (2007). *Connecting the Drops Athabasca River Expedition.* Disseminated online at http://www.connectingthedrops.ca/photos. Accessed December 18 2010.

Wong, S. & Sharp, L. (2009). Making power explicit in sustainable water innovation: re-linking subjectivity, institution and structure through environmental citizenship. *Environmental Politics, 18*(1), 37–57.

Young, I.M. (1990). *Justice and the Politics of Difference.* Princeton, NJ: Princeton University.

Chapter 8
A View from the Future

Where We Are Now

As of this writing, tar sands operators have weathered the global recession rather well, and production is barreling along anticipated growth trajectories, currently standing at plus or minus 1.5 million barrels a day, projected to reach anywhere from 3 to 5 mbd over the next decade. If anything, the recession has been greeted by local municipalities with an ironic dose of relief, offering an unplanned "pause" in the pace of development. If the record of new applications foretelling future operations is any indication, mining will soon dwindle and deep-well drilling – SAGD and in situ are the favored techniques – will become the dominant source of Athabasca bitumen. Production rates are largely dictated by incremental actions undertaken by the provincial state and corporations according to a decision cycle that is not conducive to large-scale, long-term planning: public land is leased to private corporations in individual blocks; corporations apply to the state for licenses to develop certain segments of those blocks; those applications are reviewed individually and routinely approved; the state establishes standardized royalty and tax rates that are rarely adjusted and when they are, tend to be heavily influenced by pressure from corporate interests; and the use of revenues or economic plans for the future are up to the whims of politicians subject to 5-year election cycles. Attempts to engage in a more inclusive style of decision-making, or at least exercises that give the appearance of inclusivity, have been largely unsuccessful, leaving state and industrial interests at the center of this policy-making community and all others making noise outside the closed doors of boardrooms.

Nonlocal resistance continues to grow, with a prominent role increasingly being played by retailers like the global companies Lush and Vogue Magazine. Greenpeace's mock tourism advertisement was especially creative, and You Tube videos called "Rethink Alberta" developed by a coalition of U.S. environmental organizations, designed to deter vacationing in Alberta, appeared to have some effect,

D.J. Davidson and M. Gismondi, *Challenging Legitimacy at the Precipice of Energy Calamity*, DOI 10.1007/978-1-4614-0287-9_8,
© Springer Science+Business Media, LLC 2011

all of which have raised the ire of the Alberta government.[1] Over the last 2 years, Andrew Nikoforuk's book *Tars Sands: Dirty Oil* and William Marsden's *Stupid to the Last Drop* dropped into discursive arenas like stuka dive bombers, catching the government and industry off-guard. Film festivals featured various documentaries, including *Petropolis* (2009) by Greenpeace and *Dirty Oil* (2009) by California-based Babelgum, to a fair bit of fanfare. Regional-level concern has maintained a persistent cadence (Pembina 2010). Most recently, the Parkland Institute released a rather scathing report on royalties (Boychuk 2010), offering convincing evidence that not only are royalty and tax rates set well below international industry standards, but the provincial government has also not even diligently collected from energy companies according to the current regime. Unfortunately, this report received barely a nod in the regional press. On the cultural front, Edward Burtynsky's internationally acclaimed *End of Oil* photo exhibit is appearing at the Art Gallery of Alberta[2] at the time of writing and Louis Helbig's *Beautiful Destruction*[3] images of the tar sands have been showing at the Ottawa City Hall and at various galleries in eastern Canada. Both prod public audiences to consider and discuss the previously unthinkable. Then the topic reached WikiLeaks in time for Christmas 2010, releasing a 2009 cable from a former Federal Environment Minister in the Harper Government who expressed frustration with his party's regulatory inaction after encountering international criticism of Canada's tar sands. Unfortunately, his concerns died with his recent resignation, and well before he could enact any legislative controls.[4]

The success of a movement is in many respects measured by the emergence of counter-movement activity (Meyer and Staggenborg 1996). Pro-tar sands forces have spent millions to promote this enterprise over the decades, much of it taxpayers' money, albeit these efforts are better characterized as routine legitimacy maintenance efforts, and targeted at Alberta audiences. Corporate domination of Edmonton and Calgary news media and infiltration of the provincial school curriculum have been noted elsewhere (Hodgkins 2010). More recently, however, state and industry proponents appear to have been caught off-guard by the groundswell of international opposition that seemingly appeared out of nowhere after decades of tar sands development. One sympathetic marketing professor at the University of Calgary described the government's attempt at a tar sands PR campaign as late, lame, and consequently ineffective. She characterized the oil sands industry as victim of both an unfair mugging by environmentalists and an incompetent Alberta state.[5]

State and corporate interests were unquestionably late out of the starting blocks, but start they have, with an infusion of new resources diverted specifically to counter-movement activities. Efforts have been directed at international audiences, while at

---

[1] See, e.g. http://www.cbc.ca/canada/calgary/story/2010/08/09/calgary-rethink-alberta-oilsands-tourism-poll-brit-us.html#ixzz0xRMzjIPd, accessed Dec 27 2010.

[2] http://www.edwardburtynsky.com/.

[3] http://www.louishelbig.com/tarsands.html.

[4] Norma Greenaway, "Jim Prentice in Wikileaks Release: Harper Government 'too slow' to fight dirty oil label." Edmonton Journal (22 December 2010).

[5] http://www.cbc.ca/canada/calgary/story/2010/08/05/calgary-alberta-oilsands-ad-campaign-winnipeg.html#ixzz0xRLlOJTl. Accessed Dec 21 2010.

the same time geared toward forging defense alliances with Albertans against these international forces. In August 2010, the Alberta Government announced an advertising campaign "aimed at improving the image of the oil sands industry,"[6] launched shortly after Greenpeace's unfurling of a banner on CN Tower in Calgary, the business center of Canada's oil industry, which targeted the cozy relationship between the oil sands corporations and government. Video coverage of this stunt was, as expected, disseminated on the Internet. Industry defenders have been quick to counter consumer boycott campaigns with boycotts of their own, as Alberta Enterprise Group (AEG) did in August 2010, calling on all Albertans and Canadians to boycott Gap and Levi-Strauss. (One news commentator suggested Albertans boycott California (Patels 2010)). The Government of Alberta's recent *Just the Facts* and *Oilsands 101* website education campaigns[7] are designed to "correct" the misinformation planted by protestors, including positive image galleries and video clips, many featuring industry employees, and images of tailings ponds that look more like picnic areas than waste ponds (see below). Likewise, a ramped up "Get the Real Story" Internet media campaign has been led by the Canadian Association of Petroleum Producers,[8] which combines professional TV, print, and web advertising including video clips, focusing directly on the same concerns raised by opponents, including reclamation, water use, GHG intensity, and environmental commitment, in one instance likening bitumen to peanut butter. Deep-well drilling and "clean hydrocarbon" futures have been repeatedly hailed by proponents as the green future of the industry, at once implicitly acknowledging the heavy environmental footprint of mining while also concealing the multiple forms of impact posed by drilling. See images of forward-looking workers and esthetically pleasing tailings ponds, featured on Government of Alberta's http://www.oilsandsalberta.ca.

These promotional campaigns have been joined by a series of proindustry books – *Ethical Oil* (Levant 2010); *Black Bonanza* (Sweeny 2010); and *The Impending World Energy Mess* (Hirsch et al. 2010) – that provide more extensive justification for tar sands development. This suite of books displays an evolution of framing tactics in response to growing opposition. Unlike earlier tropes that characterized tar sands development in terms of free markets and technological optimism, these more contemporary counter frames are more nuanced, framing the tar sands as not exactly clean but less dirty than alternatives, as a problematic but necessary development in an energy-hungry world, and for some, a means of transition to a new energy age (Fig. 8.1).

In December 2010, Canada's business newspaper, the *National Post*, declared that the fight over image was heating up, with state interests finally stepping up to the plate.[9] Extensive media coverage of a recent publication in the *Proceedings of the National Academy of Sciences* (Kelly et al. 2010), which directly challenged

---

[6] http://www.cbc.ca/canada/edmonton/story/2010/08/05/calgary-alberta-oilsands-ad-campaign-winnipeg.html. Accessed Dec 27 2010.

[7] http://www.energy.alberta.ca/OilSands/1710.asp. Accessed Dec 20 2010.

[8] http://www.capp.ca/Pages/default.aspx#D7MQCUKeT0Gb. Accessed Dec 17 2010.

[9] http://fullcomment.nationalpost.com/2010/12/02/kevin-libin-emerging-oil-sands-ad-battle-getting-personal/.

ENVIRONMENTAL
CHALLENGES AND
PROGRESS IN
CANADA'S
OIL SANDS

Fig. 8.1 Cover of recent report by Canadian Petroleum Producers

state and corporate claims of "naturally occurring" toxins in the river, raised the bar another notch. This time the federal and provincial states both responded, with parallel independent reviews of the adequacy of environmental monitoring. The federal review resulted in findings that could hardly have been more incriminating, exposing the shortcomings of the current program that relies on industrial self-monitoring. While this discursive contest has undoubtedly reached the ears of workers in the tar sands, none of this activity has had obvious effect on development as of yet; the Athabasca landscape continues to be dug, drilled, paved, and pipelined at an ever-growing pace.

The Precipice

Where are we now? We stand on a precipice. Earth is host to a population of 6.8 billion and growing, all of whom need resources. Not just energy, but food, water, timber, and a myriad of rare and not-so-rare minerals, as well as living wages, health care, education, and so on. While there is absolutely no doubt that current consumption of staples is woefully inequitable, and many of us can do with far, far less, the notion that all or any portion of this population can somehow become "de-materialized" is pure human exemptionalism. These material inputs to civilization must come from somewhere. Those somewheres are increasingly remote, large in scale, and associated with increasing levels of geographic challenges and "messy" sociopolitical impacts. Meanwhile, climate change has bested all predictive models, with rates of warming and melting surpassing even the worst of worse case scenarios. In short, we have a problem, and those favored problem-solving strategies that have been so fruitful in the past – mainly investing loads of new energy into new, fancier machines and institutions (Homer-Dixon 2006) – will only make things worse.

On one side of this precipice is continued reliance on those energy-intensive problem-solving strategies, meaning demand for fossil fuels will continue along its current trajectory of exponential growth, as will efforts to satiate that demand, combined with runaway climate change. On the other side is a postcarbon world, meaning a search for entirely new problem-solving strategies. The Athabasca tar sands, as the world's first large-scale industrialization of a nonconventional fuel source, marks this precipice in many ways. Can we change course? Throughout human history, societies have changed course in reflexive response to actions that are collectively defined as illegitimate. Certainly, the Renaissance describes such a collective response. The seemingly inevitable outcomes of the twentieth Century nuclear arms race were not in fact inevitable; we (actually, mostly mothers and housewives) said no. Revelations of the Holocaust could be considered a global moment of reflection that brought awareness to genocide. The tar sands, some may argue, could serve as the next Blue Planet, a new symbolic way of viewing our relationship to Earth. According to many social theorists, the photographic image of earth taken by Apollo 17 in 1972 changed forever our understanding of humanity, an icon representing "spaceship earth" – its physical limits, fragility, and global interconnectedness (Cosgrove 1994). Since Apollo's flight, this is easily one of the most disseminated pieces of visual discourse in existence (Fig. 8.2).

The visceral and interpretive impacts of this image of the Blue Planet, of the beauty of our planet and sense of global community, contrast sharply with the visceral

Fig. 8.2 This is a computer version stitched together from the following sources. NASA Goddard Space Flight Center Image by Reto Stöckli (land surface, shallow water, clouds). Enhancements by Robert Simmon (ocean color, compositing, 3D globes, animation). Data and technical support: MODIS Land Group; MODIS Science Data Support Team; MODIS Atmosphere Group; MODIS Ocean Group Additional data: USGS EROS Data Center (topography); USGS Terrestrial Remote Sensing Flagstaff Field Center (Antarctica); Defense Meteorological Satellite Program (city lights). Used with permission

impact of aerial and satellite images of the tar sands appearing on the Internet and Google Earth. The inverse of the impact of the Blue Planet, these images have far more in common with the initial impact on public consciousness of the images of Hitler's concentration camps, depicting an industrial landscape of devastation that Urry (2010) argues results from our "consuming the planet to excess," and which symbolize not hope, but human folly. While the Blue Planet spoke of *Our Home,* images of the tar sands speak of *Our Future,* one in which humans are complicit in ecological and social destruction resulting from the exploitation of the last remaining oil on the planet. Can these satellite and other images support arguments for social change? Or will we global citizens stand by and watch? Will our grandchildren look back and ask "why didn't they do anything? (Fig. 8.3)."

Fig. 8.3 Athabasca Oil Sands NASA Earth Observatory image by Jesse Allen and Robert Simmon using EO-1 ALI data courtesy of the NASA EO-1 team. Caption by Holli Riebeek. Date July 29, 2009. http://earthobservatory.nasa.gov/IOTD/view.php?id=40997. Author NASA Earth Observatory. Creative Commons. Readers can also consult the Google Earth image http://maps.google.com/?ie=UTF8&t=k&om=0&ll=57,-111.45&spn=0.415856,1.257935

Scenario 1: Well-Traveled Pathways

In Scenario 1, our collective goal is to meet projected oil demand, translating into increased reliance on nonconventional fossil fuel sources. We continue along our current development trajectory in Alberta: conventional sources of oil are depleted, and tar sands production expands to five or more mbd. In an attempt to maintain legitimacy against increased international resistance, some accommodations would need to be made, starting with amelioration of environmental impact, initially with increased investments in research and development. On the optimistic side maybe the exorbitant investments that have already been made (particularly by the provincial state) in carbon capture and storage will pay off and we will be able to sequester a significant portion of the carbon dioxide emitted by extractive and processing activities. The implementation of such technologies will be hugely expensive, driving up production costs, and thus costs to consumers. Some intensity gains are already being made in the efficient use of energy and water inputs, lowering the per barrel demand for these resources, and these could be enhanced with increased investments in research, investments that would no doubt be held up by state and corporate interests as evidence of their environmental responsibility. But in order to translate into any real net reductions in consumption of these resources, these efficiency gains would need to surpass the continued increase in production volume *and* the increased inputs needed to exploit a resource of continuously declining quality. At any rate, historic increases in efficiency have in the past only encouraged further production expansion (Bunker and Ciccantell 2005), what others call "the curse of energy efficiency" (Foster et al. 2010).

Future supplies of water are uncertain, and the state would face growing pressure from industry to increase its allowable water withdrawals from surface and underground sources. Flows in the Athabasca are already highly irregular, and climate change forecasts include increasing uncertainty of supplies, and growing regularity of drought conditions. The state would also be compelled to secure new sources of energy as the pace of development outstrips natural gas supplies. The application of cogeneration facilities that make use of industrial waste on-site can ease but not eliminate the energy demand. Eventually alternative sources would be necessary and nuclear or coal are the most likely to be seriously entertained; damming of the Athabasca for hydro is also a possibility.

The potential for reclamation, especially of tailings ponds, to anything resembling a productive landscape capable of providing some semblance of ecosystem services is highly speculative; more likely this land base would basically need to be written off as one of many costs required to meet the demand for fossil fuels. There is research being done currently that holds promise for reduced wastewater production, and thus the intensity rate of tailings ponds inflow in the future may go down. Construction of containment structures less prone to leakage is certainly within our technological grasp as well. But the need for storage of liquid waste will certainly not disappear altogether, and the legacy of the first ponds to be constructed will be with us for decades, perhaps centuries. These tailings ponds will continue to leak into the Athabasca River, and the prospect for catastrophic failure of any one of these compounds is not at all implausible, particularly under conditions of rapid climatic change, which will impose changes to the substrate in response to temperature fluctuations, and extreme weather events. Many riparian food sources would likely need to be banned from consumption, imposing upon downstream Aboriginal communities the need to import foods from elsewhere, and the increased economic and health costs of doing so. Climate change will also pose challenges to the integrity of much of the transportation infrastructure, requiring increased costs for the construction and maintenance of pipelines and highways.

The provincial state would need to divert more of its revenues away from the cities and invest into maintaining the essential services needed to support the frontier energy cities like Fort McMurray, Cold Lake, Peace River, or new urban growth nodes proposed for 200 km north of Fort McMurray. The possible costs of a future nuclear power plant (and nuclear waste disposal) would also be shifted to taxpayers. Unless some rather ingenious creative accounting can conceal such diversions, this will pose a formidable challenge to elected officials who must reckon with disgruntled urban voters who make up the majority of the electorate, and the energy corporations to which the province will inevitably turn for increased royalties and taxes in order to pay for such services. Fort McMurray will be, as is ever the case

Fig. 8.4 Municipality of Wood Buffalo Attractions webpage. "'Experience the Energy' of Fort McMurray with a tour of the Suncor Energy mine site. See the earth move before your eyes as shovels carrying 100 tons load 380 ton payload trucks with the rich, black oil sand." http://www.woodbuffalo.ab.ca/visitors/attractions/images/trucks.jpg. Used with Permission

for staples economies, subject to the whims of a fickle global marketplace, liable to see the divestment of industry as rapidly as investment.

At the global scale, if we (admittedly simplistically) assume that the ecological costs of various forms of nonconventional fuels are roughly comparable, then an ecological sacrifice zone roughly 67 times the current Athabasca footprint would need to be accommodated in order to produce the 105 mbd of oil that is projected to represent global demand by 2030, as we eventually become completely reliant on hard-to-reach and nonconventional fuels. Likely sources already attracting interest include the Venezuelan Orinoco, the Arctic sea bed, and the Caspian Sea, among others. The passage of time will be marked by movement ever-closer to that energy cliff described by Murphy and Hall, in which the costs of energy inputs to maintain our economies will eventually pose a systemic fiscal crisis. Greenhouse gas emissions will continue to grow, and worst-case climate change forecasts will be borne out, increasing further the fiscal burden on societies, while also introducing with increased regularity instances of large-scale human catastrophe, in the form of food shortages and natural disasters. Trucks filled with Athabasca tar rumble on, poetically driving us all into the sunset of modern civilization (Fig. 8.4).

There is no shortage of indicators that some variation of Scenario 1 will be borne out. Global resource accounting complies with Peak Oil projections; with new discoveries

Renewables Are Flatlined

The total proportion of energy provided by renewable energy sources has remained constant (actually it has declined slightly) since 1973. In 2008, it was 10% (International Energy Agency 2010). The investments that are taking place, furthermore, tend to be in large-scale centralized projects that are suited to industrialized societies and beyond the reach of many less-developed countries, where small-scale technologies that could be implemented at the household level could make a tremendous difference in the well-being of millions of families.

ever smaller and more challenging to access. States and corporations alike are clamoring for control over these remains, with the few jurisdictions seriously contemplating alternatives as the exceptions that seem to prove the rule. Enthusiasm for international environmental governance is waning; efforts to establish even the most timid of international climate change protocols appear dead in the water. Back home in Alberta, the Progressive Conservative Party is suggesting more of the same prodevelopment policies as the previous decade. The Alberta Government's Oil Sands Secretariat's Responsible Actions: A Plan for Alberta's Oil Sands, and C.R.I.S.P., a comprehensive regional infrastructure sustainability plan, are 20- and 30-year plans that anticipate full-steam ahead growth, with proposals on how to mitigate every socioeconomic and environmental fallout. The Party's most formidable challenger is the up-and-coming Wild Rose Party, sitting further to the right of the political spectrum. On a deeper level, as the just in time "oil sands development sustainability plans" indicate, the human exemptionalism paradigm, which prescribes faith in future economic and technological solutions to ecological travesties committed today, is being contested but by no means replaced.

Scenario 2: Transitions to Resilience

Scenario 2 must begin with a clarification of terms, the need for which exemplifies just how much interpretive frames really matter. We employ the term 'Transition' to denote passage from one institutional state of existence to another. It is far less prescriptive than other terms in similar use, such as revolution, as it does not forecast the particular form in which change processes might take. However, it should be noted that this term has become increasingly popular not just among those promoting a postcarbon economy, but also among those in full support of continued fossil fuel development. As such, we distinguish those changes proposed by fossil fuel proponents as small-t transitions, while movement toward a postcarbon economy is best captured with a capital T – Transitions. Scenario 1 embodies several small-t transitions as needed to ensure a long-term plateau of energy production: technological solutions to mitigation, reclamation, and other problems associated with exploiting nonconventional fossil fuels; and maintenance of the transportation and industrial infrastructures of the current global economic system. The Alberta government's recent depiction of a "clean hydrocarbons" energy future is emblematic of just such a focus, with plans for increased in situ extraction of oil, GHG intensity targets, water conservation, and investments in reclamation research, combined in a bundle of discourses to promote not so much "business as usual," but rather an embrace of the formidable challenges ahead with the tried and true employment of technological ingenuity.[10]

Scenario 2, on the other hand, implies a big-T Transition, demanding change across the industrial, economic, political, technological, and cultural institutions characterizing contemporary society. As such, it is somewhat more challenging to portray, as it departs so dramatically from current social trends. Visions of alternative futures certainly abound, and we do not intend to inject yet another into the literature here. Instead, we begin with the ends rather than the means: the foundational premise that greenhouse gas emissions, and by extension fossil fuel consumption, would need to be reduced drastically: let's say 90%. Globally, state and corporate investments in exploration and development of fossil fuels would necessarily be diverted into conservation and alternative energy development, including massive investments in the physical infrastructures that would be needed to change our current energy provision network. In Alberta, a moratorium would be imposed by the state (Canada and Alberta) on further tar sands development, and current operations would eventually be phased out, the interim surplus value invested into environmental amelioration and economic diversification. And underlying all such shifts, the *need* for change must first be embraced. We will tackle this condition momentarily, but first, let us continue with our scenario-gazing.

Scenario 2, often depicted in utopian terms by visionaries, would be far from it. Energy, to put it mildly, will be expensive, very expensive, and in short supply. This means fuel poverty for a large proportion of the global population and an industrial enterprise that would be brought virtually to a standstill, particularly the heavy industries. Political pressures from irate businesses, consumers, and citizens would pose a formidable legitimacy challenge to any and all states that did not have a solid Transition plan in place. No matter how much of an optimist one is, this is not a rosy picture. Rather, in contradistinction to Scenario 1, it is a picture of survival, and represents the only viable route away from the energy and climate cliffs that are the inevitable final destination of Scenario 1. There are certainly some upsides – many in the West may well find their quality of life enhanced in numerous ways once relieved of the never-ending race to material betterment. Walking and cycling are healthier than the use of private automobiles, both physically and socially. Community gardens improve the esthetics of neighborhoods, and fresh food is always preferable. We might get to know our neighbors better, with benefits for social capital. A renewed enthusiasm for political leadership and participation may

---

[10] See "Alberta's Energy Future," available at http://www.energy.alberta.ca/Org/pdfs/AB_ ProvincialEnergyStrategy.pdf. Accessed Dec 27 2010.

unfold, with citizens everywhere motivated to become engaged. But economic growth would falter to say the least, and regional economies dependent upon fossil fuels would suffer more than most. The lowered emissions would eventually stabilize and then reduce the proportion of greenhouse gases in the atmosphere, but the impacts of climate change in response to past emissions would continue to unfold for some time. Mobility of people and products would decline, particularly air travel.

Relocalization of food and energy production would occur, which may not hamper some regions, but one must remember that supplies of fertile soil and energy are not equitably distributed. Central California and Italy may fare well. Communities in sub-Saharan Africa, the southwestern U.S. and northern Mexico, interior China, and northern Scandinavia, not so much, and this implies the need to continue to redistribute food and fuel, face the mass migration of entire regions, or both. Even in those regions still blessed with fertile soils, as a result of the mass urbanization of global populations over the past 50 years, both access to land and skills to produce food are disappearing. Global food production levels would drop below those implied by Scenario 1, at least in the short term, since we would no longer have fossil fuels available to support the intensification of agriculture, which, combined with the distribution quandaries just mentioned, implies the need for greater political intervention in agricultural markets to ensure equitable distribution of food. Developing economies in the South, with burgeoning populations of impoverished families, would continue to suffer, although one could argue they wouldn't necessarily be any *worse* off than they would be in Scenario 1, and very likely they could be better off in the long run. For one thing, global capital would not be scavenging their territories for resources at the same rate, and the increased availability of renewable energy technologies that are more applicable at a small scale than are fossil fuel-based supplies might well *increase* the energy availability in many less-developed regions.

Even with unified support for Transition, moreover, the reality is that the tar sands are not going to stop being developed tomorrow, and our global economies are not going to suddenly stop consuming fossil fuels. This Transition does not imply an abrupt shift between two distinct phases of modernity. It suggests a powering down of fossil fuel use combined with a simultaneous powering up of new industries that are far less energy consumptive, and alternative forms of energy, state and social configurations, forging a long and rocky shift.

Getting There: Prospects for Scenario 2

Okay, not so rosy, but realistic and certainly preferable to mass suicide, and many, many upsides, some of which will likely only reveal themselves in process. Stargazing is always a fruitful intellectual exercise, but the question we would like to devote much of the remainder of this chapter to is the question of plausibility. First, what are the conditions that would need to be in place for such a Transition to occur, and secondly, do we see signs of those conditions emerging? *Some* degree of

"I THINK YOU SHOULD BE
MORE EXPLICIT HERE IN STEP TWO."

change may well happen in what appears to be a more or less spontaneous, or at least unplanned, manner. The increased input costs for oil production are likely to eventually make this investment less attractive to capitalists in comparison to other options. Lowering aggregate household energy consumption could conceivably be achieved via the aggregation of individual behavioral change alone. A massive drought in Alberta would shut down tar sands production in a heartbeat. But all things considered, a large-scale Transition to a postcarbon society will demand a form and level of collective agency that may well be historically unprecedented. As has been discussed throughout the book, such a change would necessarily begin with a large-scale legitimacy crisis, borne by the mobilization of a distinctly global and environmental citizenship that is premised on responsibility as much as rights, which in turn is premised on a questioning of Western ideology – a tall order to say the least. So, fossil fuel dependence must first be defined as unacceptable, particularly among those who are consuming the largest proportion of them. Then, change would require the imagining of life without it, an even taller order. And finally, societies must exhibit the capacity for international-scale, goal-oriented collective action – in effect *intentional* social evolution – which amounts to the tallest order of them all (Fig. 8.5).

Can this be done? We won't know for sure until it happens (or doesn't), but we posit that it is at least theoretically possible. Contemplating this plausibility requires some hard thinking about just how social change does occur. Unfortunately, many treatments of macrosocial change, scholarly and otherwise, have difficulty problematizing those critical "step two's," the mediating processes that define the causal relations between conditions and responses. Reflexive Modernization (e.g., Beck 2009; Beck et al. 1994), to take one pertinent example, postulates that citizens will become alarmed at the risks associated with modernity, mobilize through various sub-political avenues, and presto, global institutions become transformed in a manner that fosters greater reflexivity as a result. Such a set of postulates is extraordinarily attractive in its functionality and uniformity; however, neither the processes by which individuals (differentially) reach awareness and (differentially) act upon that awareness, nor the means by which such action will cumulatively result in the (differential) transformation of global institutions are problematized. In short, there is no guarantee that the Athabasca tar sands or our impending future of energy calamity will be interpreted as a crisis by all, nor that all such interpreters will engage in ecologically minded reflexivity as a result, or that such efforts where they do emerge will enable a reasonably civil transition to a postcarbon economy. We do find some encouraging signs observed in the current study, although our optimism is cautious.

Reflexive modernization essentially assigns agency to institutions, while the role of individuals – the only social entity capable of agency, reflexivity, and the political reconstitution of institutions – is obscured. Archer (2003; 2007) offers a useful corrective to Reflexive Modernization Theory by bringing attention back to the individual, and as a result, offers a much more developed theorization of reflexivity than does Beck and his colleagues. Reflexivity describes "the regular exercise of the mental ability, shared by all normal people, to consider themselves in relation to their contexts and vice versa" (Archer 2007:4). Archer offers a compelling research record to support the position that human behavior and thinking is neither deterministic nor directed solely by socializing conditions, but rather is mediated by internal conversations (Archer 2003), which guide the propensity for problematization of current conditions, imagination of alternatives, and activation of behaviors intended to promote change. Structures present forces of enablement and constraint to individuals, and individuals interpret and negotiate those forces on the basis of their own personal projects, defined by unique sets of values, experiences, and goals.

But individuals must also *choose* to pursue certain projects from among a universe of imagined possibilities, and this is where structure matters. The extent to which a chosen pathway has been traveled is an inverse measure of reflexivity. Doing things the way they have always been done according to social memory implies a well-traveled pathway and relative lack of reflexivity. Our imagining a future without tar sands development in Alberta, or without fossil fuels in the global economy, demands a particularly high level of reflexivity. Well-traveled pathways also have deep ruts, making it even more difficult to change direction, suggesting that even if an alternative pathway can be imagined, steering out of the ruts in order to embark on a new direction is another thing entirely. The social organization of consumption, for example, is deeply entrenched in the West, such that purportedly individualized material preferences and consumption patterns reflect everyday infrastructures and technologies such as building design and material use, transport systems, and urban planning (Shove and Walker 2010). Even preferences for bathing and household comfort practices are not as individual as they seem. According to Shove and Walker, they are premised on social expectations, first of all, but physical infrastructure matters as well. In fact, the infiltration of domestic appliances goes a long way to *create* those expectations (there is no excuse for wearing a dirty shirt when everyone has a clothes washer). These organizational precursors to everyday practices, and the imagination of alternatives, are invisible to most, many such practices being forms of habit more than conscious agency.

Globalization has engendered one set of particularly imposing structures onto individual decision-makers with significant influence on prospects of reflexivity, although this influence is a bit of a double-edged sword. The complexity of our current global system is characterized by fast reactions, multiple unintended pathways, and unpredictable, latent effects that, on the one hand, present foreboding possibilities for reflexivity. Thomas Homer-Dixon (2006) said it best: "As we aggregate ourselves into larger and larger groups, we seem to behave more and more stupidly." This may in large part be due to the fact that global complexity supports a relinquishment of responsibility for the consequences of individual or organizational behavior. The sheer increase in the number and frequency of independent transactions promotes a sense of disempowerment (I am just one among 6.8 billion, changing my behavior won't make any difference). Even for those without such fatalist tendencies, our cognitive limitations in the ability to conceive the higher-order effects of our actions within such a complex system constrain forethought. Many actions at the microscale may be perfectly reasonable in their own right – the selection of lowest-cost production inputs; refusal to report certain observations for fear of getting fired; the decision not to invest bureaucratic resources in changing entrenched policy regimes; and the use of private automobiles to save time in busy households. And yet in the aggregate, such perfectly rational microscale decisions amount not to betterment for all; such rationality at the microscale culminates in *ir*rationality at the macroscale.

On the other hand, globalization simultaneously imposes the need for *increased* reflexivity onto individuals, as the structures that previously represented them – state, class, gender – have weakened (Bauman 2000). The system of scapes and flows characterizing contemporary global societies also offers opportunities for reflexivity, exposing individuals to new conditions and information that can challenge accepted meanings, raise new concerns, enable the imagining of alternative futures both good and bad, and also introduce new avenues for collective action. The very uncertainty/crisis tendencies associated with the intensification of transactions in complex societies means that the utility of certain habits, beliefs, and routines are more often challenged: doing things "the way we always do things" just doesn't get us to the same end point anymore. Such shocks are opportunities for reflexivity.

The real challenge is the transference of this individual reflexivity into collective action, and subsequent institutional change. Many human projects are individual – finding a mate, paying off debt, avoiding toxicity – but many, many others must necessarily involve collective engagement, meaning that their pursuit demands collective action that culminates in political influence. How does individual reflexivity translate into collective reflexivity? Those individual projects must first and foremost encompass a desire to confront social institutions, and in this case, that institution that must be confronted is the state. We do not disagree with many recent critiques of the increasing revelations of ineptitude expressed by our nation-state system when it comes to dealing with global problems. Nor do we disagree with assessments of the frustrating rigidity observed in centralized bureaucratic institutions. Sometimes the rigidity of state regulatory structures not only allows unabated pollution and land disruption, it can even serve as a barrier to corporations seeking to capitalize on those shifting social goal posts, by preventing ingenuity.

But states are nonetheless essential, not only because nearly every deposit of fossil fuel across the globe is under the formal authority of one state or another (and some are being argued over, such as those above the Arctic Circle). Also because, when it comes to most supplies of natural resources, the real power holder is quite explicitly the state; indeed, control over resources such as timber has been a primary role of states for as long as states have been in existence (Perlin 1989). States are also essential players in Transition because no other social institution has the coordination capacity that states enjoy. As Tony Judt argues, the state remains the best and only tool we have at hand with the scale and range of powers to carry out the collective changes that a big-T Transition requires (Judt 2010:175–200). And some states have shown us that reflexive state-making is possible. Consider the Norwegian state and its use of oil revenues to plan for the end of oil, to take just one pertinent example. Many states, and arguably the very Westphalian nation-state system, are broken, but the *idea* of a state system, as a set of coordinative authority structures established for the purpose of problem-solving, is not. To the contrary, such a set of structures is at least as essential today as it was at the birth of the state system some 500 years ago. Thus, our only recourse is to fix them.

Signs

For the legitimacy of our current relationship to energy to be challenged, the invisibility of this relationship must first be made visible (Shove 1997). Raw materials are invisible to us in many ways: the companies that produce them, the ships and trains that transport them, and their fundamental contributions to the commodities and uses to which we put them are all inconspicuous, hovering below the surface of our attention (Shove and Warde 1998). Energy is perhaps the most invisible of all. It is quite literally invisible in many ways; our consumption of it in homes and cars only detected by means of meters. And it is invisible in its very ubiquity, because energy is either created or, far more often, consumed, in virtually everything we do, practices that are rarely subject to question (Shove and Walker 2010). And finally, the current generation of Westerners has rarely, if ever, experienced the *lack* of energy.

Yet energy is indeed, in fits and starts, becoming more visible. Analysts of EROI offer one attempt to make the invisible visible, highlighting the kinds and amounts of fuels used to produce other forms of fuel as a way of increasing the transparency of our relationship to energy. Images of tar sands operations offer another. The increased incidence of natural disasters that leave communities without power is a third. Each of these provide larder for those internal conversations identified by Archer, planting reflexive seeds of doubt, or at least contemplation. The key means by which the invisible and inconspicuous are rendered visible, the means by which the immobilized become engaged, the means by which states are subject to legitimacy challenge, and the means by which organized collective action is enabled is through discourse. While Archer focuses most of her attention on the internal conversations entertained by individuals, and Shove on social structures, discourse defines the link between the two, as the means by which agents receive inputs critical to their internal conversation, and hence into practices, and ultimately voice new demands. Discourse can encourage or discourage interpretations of our relationship to energy as a crisis and can offer imagined alternatives (superindustrialization vs. alternative energies and conservation). Discourse is the means by which habits, practices, values, worldviews, and models of legitimate governance are nurtured or subject to challenge. Agents may embrace or reject that scrutiny depending on one's own reflexive predispositions. On the other hand, the absence of scrutinizing discourse means those engaged in reflexive internal conversations have less to work with.

The first place to look for signs of change, then, is in discourse. We have offered evidence throughout this book of such signs, particularly in the narratives offered by local citizens. Those concerned Albertan citizens who have been motivated to engage politically represent a small minority of the populace, but what they have to say is quite intriguing. Citizens who have testified at the Oil Sands Consultations have offered narratives that bring visibility to many of the contradictions of our oil-dependent system with a level of poignancy that is only possible with first-hand experience and threats to home and health, and they have inserted more explicit challenges to the provincial state itself. And while the number of testifiers is small, survey research indicates that the proportion of Albertans who have not quite accepted the state's narrative lock, stock, and barrel is actually rather large. Resistance among municipal elected officials offers a powerful ally for concerned citizens, as have scientists and other experts who have validated claims of environmental and health consequence. When we include Canadians from other provinces, who are less inclined than Albertans to feel as if they are betraying their heritage, or their wallets, resistance is even larger. In Canada and Alberta, the protestor discourse that has emerged has uniformly called for a moratorium on expansion and supports the assertion of state management of the pace of development, and its social and environmental impacts. This resistance may be aided by growing splits among ruling elites in Alberta (by generation and by ideology) and these splits will only deepen with ensuing raw materials supply crises: forestry is already fighting with oil and gas over land; oil and gas will soon be fighting with agriculture over water; and everybody will be fighting over dwindling natural gas supplies.

These seeds of doubt will only lead to Transition if planted into states; however, beginning, in our case, with the Provincial state of Alberta, and it is to this prospect we now turn.

But What About Capital?

Some will immediately point to the seemingly invincible power of multinational corporations to manipulate the hands of government actors and throw up their hands at the prospect for state engagement in Transition. States depend on economic development to fill their coffers and employ their citizens, and these prerogatives underlie much state activity regardless of the presence of political pressure from capitalists. And while capitalists may encompass all creeds (including the socially and environmentally conscious), regardless of the personal predilections of CEOs, board members, and stock holders, corporate behavior is ultimately defined by the bottom line. This bottom line means, as we know, that costs will be externalized whenever possible, market expansion will always be a priority, and if it doesn't have a dollar value, it is not likely to be factored into decision-making. It also means that the business elite will be motivated to pressure the state to facilitate the maximization of private profit, by enhancing market access through the negotiation of trade agreements, ensuring a skilled labor pool at low cost, providing access to raw materials, and so on. Their effectiveness in doing so has crippled many weak states and has elevated the level of turbulence in what are already very dynamic and complex global economic systems.

These features of industrial capitalism have fostered the overconsumption and waste that are now coming to roost, and yet these same imperatives make capital extraordinarily practical – reflexivity is elemental to survival in a global economy. So, if the Alberta State was to impose constraints on tar sands development, a capital strike by global corporations would most certainly be predictable, but shifts in the goal posts instilled by strong states, backed by strong civil societies, are ultimately more likely to simply be met with new corporate game plans. Corporations routinely engage in future planning with 25-, even 50-year time horizons (compared to the 2–4-year time horizons of most elected officials), in which all possible scenarios, however unpalatable, are considered. Shell, one of the world's largest energy corporations, has a 2050 future plan in which peak oil, scarcity, increased demand, even public ownership are tossed onto the vision board (Zalik 2010) – in short, capitalists cannot afford invisibility, anymore than they can afford to spend limited resources resisting change. Freed from the normative imperative of determining whether such potential outcomes are ideologically compatible, corporate planning exercises revolve solely around the question "how can we make money under this scenario?" So while it may be considered prudent corporate policy to invest in campaigns that ensure maximum privatization of profits and subsidization of costs when such tactics are perceived to have some probability of success, corporations like Shell are already planning for survival in the face of possible variations of Scenario 2.

Corporations operating in the tar sands have polluted, exploited, and made spectacular sums of money. But critics would do well to acknowledge that this is exactly what is to be expected, because these are the sorts of activities that have been *enabled by the social and political structures we have created*. And as is equally expected, corporations have sought to keep it that way, expressing strong resistance to increased royalties and environmental regulation, seeking the cheapest means of securing labor possible, and making persuasive arguments regarding their lack of culpability for everything from climate change to employee housing, all the while insisting in discourse on the legitimacy of their behaviors. But ultimately, capitalists are quite possibly the most reflexive members of society today, and if and when they confront social and political structures that offer more incentives for change than for resistance, then they will change. Those social goal posts are already beginning to change, with consumers demanding greener, ethically produced products in greater and greater numbers, and corporations in many sectors appear to have been more accepting of, and adaptable to, these new goal posts than have our regulatory structures.

The State Has Mind(s) of Its Own

Many observers attribute the Albertan State's adamant support for the tar sands to the influence of energy corporations. But it would be more accurate to say that energy corporations are acting in the state's interest in this instance, albeit with mutual benefit. Lest we forget, this small provincial state *has* exhibited an extraordinary degree of autonomy in the past and perseverance in the face of miserable odds and repeated failures. The current Premier takes every opportunity to remind listeners of the determined, innovative spirit of Albertans, and despite the many, many topics on which we wholeheartedly disagree with Premier Stelmach, on this particular topic he is right. It was just such stubborn determination that, over the course of five (!) decades, allowed for the commercial-scale production of bitumen, decades during which private capital only played bit parts. It is this very determination, or more accurately the fruits of that determination, which may well be blinding elected officials and their bureaucratic appointees to the need for consideration of alternative pathways. "We have worked so hard for this, dammit, we aren't going to let a few environmentalists (and energy analysts, and physicians, and university scientists, and municipalities, and opposition parties) get in our way!" But this very history, oddly enough, does provide a rather compelling counterpoint: if the Albertan state apparatus could so diligently see tar sands development to fruition despite the naysayers, it can see other challenging visions to fruition as well, including a Transition to a postcarbon economy. How is it that representatives of that same apparatus who celebrate Alberta's can-do attitude can turn around and say "we can't" to alternatives? The Albertan state would appear to be perfectly capable of engaging in the promotion of a Transition to a post-tar economy, perhaps more capable than most if history is any indication.

But state actors would first need to be motivated to do so, and that is unlikely to transpire without a full-scale electoral upheaval in the legislature at the very least. The political terrain in the province, if anything, has veered further to the right, with the PC Party gaining seats in the last election. Although the Party's steadfast ascription not just to tar sands development, but to neo-liberalism, may contribute its eventual undoing, Alberta's behavior over the past decade (no planning, minimal environmental safeguards, low royalties) has created impacts so blatant that they are becoming more and more difficult for proponents to conceal or explain away, particularly in the Information Age, in which visual testimonials have flourished. One need only look next door to the Province of Saskatchewan for signals that enthusiasm for neo-liberalism is waning. The proposed sale of Potash Corporation of Saskatchewan to BHP Billiton met extensive resistance from many corners of Western Canada, in order to protect "a vital resource" that serves the interests of citizens[11]; yet another instance in which citizens called on the state to intervene in the marketplace to ensure protection of the public interest.

These emerging contradictions, cataloged throughout earlier chapters, have been the impetus for growing opposition – from municipalities, from Liberal and New Democrat MLAs, from scientists, from citizens, from movement organizations. The federal state should not be discounted either; having been subject to those same pressures, the feds are beginning to exert some muscle in the province. While Calgarian Prime Minister Harper has been extremely hesitant to infringe on tar sands development, growing pressure from many fronts is beginning to be felt, and there are many representatives of the national state who are far less beholden to the tar sands industry than is the current Premier. A contingent of Senators recently visited Fort McMurray and offered colorful personal accounts to the media that departed in no uncertain terms with the glowing reports offered by Harper, Stelmach, and others. The Canadian state has also most recently exercised some authority by striking an independent scientific review of the environmental monitoring programs in place in northeast Alberta, which resulted in a scathing condemnation of the provincial state's regulatory record.

Such recent events offer opportunity windows, but do not of themselves provide sufficient impetus for a Transition. Energy is becoming visible, but the state's culpability in the management of society's relationship to energy, and its necessary role in an energy Transition, is not yet as visible as it needs to be. The presumed "retreat of the state" by many political actors and theorists has inspired a shift in much social movement organization targeting away from states and toward markets and cultural institutions. This has broadened the scope of movement activity, with many positive benefits. It has also meant relatively less challenge by social movement organizations which, if they engage with the state at all, are more likely to do so under the conciliatory and nonconfrontational terms of inclusive governance – we may get higher fuel efficiency standards from such efforts, but not a challenge to growth. Most tragically, this retreat of the state perspective has transpired into the disengagement of citizens in state activities. What has rendered the state a broken entity more than any other condition is the disengagement of the citizenry. Certain state actors, recognizing the freedom that citizen disengagement offers, have done little to remedy this situation, and some have further promoted this retreat narrative – it is, after all, the crux of neo-liberalism. Premier Klein and those who followed him have

---

[11] http://www.theglobeandmail.com/news/politics/hometown-opposition-to-potash-takeover-puts-harper-in-bind/article1782642/.

worked hard to remove the Alberta state from public service provision, championing privatization of public services and free market principles, while appeasing citizens with various participatory processes: stakeholder groups, public hearings, advisory boards, and the like. All the while the provincial state has continued to use their substantial state power to manipulate royalties, taxes, land concessions, labor laws, environmental policy and regulations to facilitate the interests of capital, all with limited oversight from citizens. We join historian Tony Judt (2010) and others in calling for liberation from this free market ideology and its negative myths about the role of the state. The state was, after all, an integral player in Western societies' political, social, and economic achievements over the last century and must be reinserted into public discussions about our future.

Reharnessing the Locomotive

Transition calls for a full-scale reempowerment of citizens to engage with their states. Citizens closest to the tar sands have already been attempting to do just that, calling not for consumer boycotts, but a moratorium on tar sands development; demanding more, not less, state-led governance; and for a reprioritization of state roles, away from endless growth – ecologically modernized or otherwise – and toward moral reckoning that privileges the families and futures of energy workers and not corporate shareholders and CEOs, and the livelihoods of Aboriginal peoples, future generations, and the poor.

Citizen empowerment can happen, but it will require the intersection of many discourses and many organizational efforts. Environmental movement organizations have without question played a critical role in supporting resistance. Local organizations like the Pembina Institute, the Environmental Law Center, and the Parkland Institute have been instrumental in offering critiques of state and corporate narratives for many years. Nonlocal organizations have been drawn to the tar sands far more recently, but have been no less instrumental, bringing visibility of the tar sands to international publics and facilitating protest activities in multiple nonlocal settings, localizing the nonlocal. In some ways, international organizations such as Greenpeace have strengths in those areas where regional organizations are weak. They have a much broader and more diverse universe from which to draw funding and sympathy. But they also face a much broader suite of agenda items, the selection of which go a long way in determining just how much of that support base they are able to capture. In all, the contributions to discourse made by environmental organizations have aided in visibility, and in many ways challenge the human exemptionalism paradigm, but as of yet have not offered narratives that have the greatest potential to generate a legitimacy crisis on their own.

Labor poses as a powerful ally for whichever side it chooses to support. While historically the labor supporting industry in rural staples-based economies has traditionally abided by corporate interests, the nature of the global labor pool supporting Fort McMurray presents a new structural relationship with capital that may lead to greater instances of political engagement by labor. The relative power of labor in the tar sands is evidenced most readily in the extraordinarily high wages being offered. This is the outcome of a tight labor market; considering the large scale and skilled labor-intensive nature of this type of development, the tar sands enterprise quickly outstripped the local labor pool and resorted to an increasingly global flow of skilled and unskilled labor. This labor mobility introduces a source of complexity and hence uncertainty in political power structures. On the one hand, the multiple backgrounds and languages present among workers limit opportunities for collective action. On the other, a larger proportion of workers will be less inclined to feel beholden to a particular employer; on the contrary, many would be on the next plane out of town were employment opportunities to emerge closer to their homes. Barring those opportunities, national and international labor have a global industrial enterprise within which to market themselves, and those energy companies working in the Athabasca must compete with labor-demanding capital everywhere. And the social networks workers are members of and not beholden to the tar sands either, which could present sources of sanction from peers to individual workers as the international reputation of the tar sands enterprise continues to be tarnished.

Individual workers have illustrated hesitancy in playing the role of whistleblower, and they have not had a strong presence in political theaters like the Oil Sands Consultation (although they have not been completely absent either). On the other hand, because much of the labor demanded is skilled, a relatively high proportion of workers are educated and computer literate. Some workers have contributed directly to political discourse through blogs and other digital avenues of communication with compelling first-hand accounts. The recent firing of that same employee was made just as public as his testimonials, causing that company's attempts at silencing its workforce to backfire. Most compelling is the position taken by organized labor. The Communications, Energy and Paperworkers Union of Canada (CEP) has released a bold statement supporting Transition, emphasizing the need not only for structural movement away from fossil fuels entirely, but also the need to ensure the justness of that transition, by safeguarding the job security of energy workers through retraining. The CEP also, notably, calls for a strong state-labor alliance. The CEP is not alone in their positioning either; labor organizations across the West are beginning to organize around green jobs master frames which open doors for coalition not only among labor organizations, but also with social and environmental movement organizations, green industries, and most importantly, states.

Aboriginal people in other regions of Canada have also become involved in the promotion of a green jobs coalition, although to date the Aboriginal communities in the Athabasca region have not done so. As elsewhere, the Aboriginal communities located in the region are plagued with a number of livelihood threats and a political history that has compromised their social capital, both of which limit their political capacity, and their trust of state institutions. Many members of these communities, moreover, despite the travesties imposed upon them by the industrialization of their homeland, would readily take employment in the tar sands as one of very few means of survival in their homeland.

However, several factors also serve to raise the specter of Aboriginal people playing a formidable role in inserting public interests into state agendas, including in particular human rights, and the discourses they have offered are quite complementary to a discourse of Transition. First, they have brought an environmental justice frame into tar sands discourse, which can be a powerful means of eliciting support, and of compelling a response by the state. European publics in particular have expressed heightened sympathy for the plight of North American Aboriginal people over the years, and while not used to such global-scale effect to date, an ecological justice frame may well be a key motivation among nonlocals to support those organizations mobilizing against the tar sands, and perhaps become mobilized themselves. The plight of Aboriginal people has been conveyed with the help of traditional knowledge as well – unlike most non-Aboriginal northern residents, Aboriginal land users have well-developed capacities to observe changes in their ecosystems and articulate these changes politically. They are, more than any other group, the eyes and ears of civil society, bringing news of the impacts of energy development to global discourse. First Nations and Métis are also well-organized and can and have brought that organizational capital to bear. Finally, they have a fundamentally different relationship with the federal (and provincial) state, one which can be employed to compel the state into action in ways that other citizens cannot.

Eight Chapters in Two Paragraphs

There are indeed signs that the tar sands has brought a heightened level of visibility, not just to the Athabasca region, but to the costs of society's dependence on oil. We can also observe growing unity of opposition to neo-liberalism – this was certainly not initiated in Alberta, but the fact that it is beginning to face concerted challenge in one of its regional strongholds is telling. Support for human exemptionalism is still strong, but tar sands development has certainly revealed many of the contradictions in this ideology. Those contradictions have been particularly notable in relation to reclamation – no one really believes anymore that ecosystems can just be recreated after we drain them of economically valued resources. No one group is in a position to single-handedly foment a moratorium on the tar sands, but in combination, there is strong potential for success, especially among labor, concerned local citizens, social movement organizations, Aboriginal people, energy analysts and ecologists, and some public sector workers within states. Global civil society needs to focus on empowering concerned Albertan citizens to confront their state, and by doing so give citizens everywhere the encouragement to take back their own states. This can happen first by aligning oppositional discursive frames with those used by citizens and thereby elevating those frames beyond local political theaters, and second by bringing the extensive nonlocal organizational capacity available in global civil society to the service of local citizens in Alberta (and elsewhere), and also by targeting directly the Alberta state. And finally, there is tremendous coalition potential under the master umbrella frame of Transition, a master frame that must include realistic plans for implementation.

Directing such global resources to bringing attention to the tar sands, some may argue, may stop production in the Athabasca, but merely increase pressure on other nonconventional fuel deposits. This may be true. Or maybe not. Shutting down the tar sands might be an isolated incidence, like protecting an old growth grove while forests elsewhere succumb to the axe, or it might just mark the start of a Transition. Much will depend on the character of that master frame proposed above. The world is watching what is happening in the tar sands, and a moratorium here in the name of postoil futures, combined with successes in local-level Transition efforts already being undertaken everywhere, could send a signal to global capital, other states, and other citizens, regarding the political, social, and environmental costs of nonconventional fuel development, and of the possibility for change. In our lifetimes, citizens everywhere have experienced transitions. Here at home, Canadians have seen the collapse of the Atlantic fisheries and many nonrenewable resources, thousands of boom-turned-ghost towns, the disappearance of the family farm, not to mention multiple natural and social disasters from which families and communities have recovered. These experiences are available to us as feedback for rational reflection and should provide a strong basis for considering the end of oil, the imagination of alternative futures, and the courage for Transition. The human species is simultaneously extraordinarily resistant to change as a result of our institutions, and yet the most adaptable species on the planet. Let's just get on with it.

References

Archer, M.S. (2003). *Structure, Agency and the Internal Conversation*. Cambridge: Cambridge University Press.

Archer, M.S. (2007). Making our way through the world: human reflexivity and social mobility. Cambridge, U.K.: Cambridge University Press.

Beck, U. (2009). *World at Risk*. Cambridge: Polity Press.

Beck, U., Giddens, A. & Lash, S. (eds). (1994). *Reflexive Modernization*. Cambridge: Polity Press.

Bellamy Foster, J., Clark, B. & York, R. (2010) Capitalism and the curse of energy efficiency: the return of the Jevons paradox. *Monthly Review, 62*, 6.

Boychuk, R. (2010). Misplaced Generosity: Extraordinary Profits in Alberta's Oil and Gas Industry. Parkland Institute, Edmonton. Available at: http://parklandinstitute.ca/downloads/reports/MisplacedGenerosity-Web.pdf. Accessed Dec 20 2010.

Bunker, S.G. & Ciccantell, P.S. (2005). Globalization and the Race for Resources. Baltimore: Johns Hopkins University.

Cosgrove, D. (1994). Contested global visions: one-world, whole-earth, and the Apollo space photographs. *Annals of the Association of American Geographers 84*, 270–94.

Hirsch, R., Bedzek, R. & Wendling, R. (2010). *The Impending World Energy Mess: What It Is And What It Means To You!* Burlington: Apogee Prime.

Hodgkins, A. (2010). Educating for Democratic Citizenship in an Oil Dependent Economy Conference Presentation Unwrap the Research Fort McMurray. Available at: http://www.uofaweb.ualberta.ca/crsc/pdfs/Unwrap_-_Educating_for_Democratic_Citizenship_in_an_Oil-dependant_Economy.pdf. Accessed 27 Dec 2010.

Homer-Dixon, T. (2006). *The Upside of Down: Catastrophe, Creativity, and the Renewal of Civilization*. Washington DC: Island Press.

International Energy Agency (2010). Key World Energy Statistics. Available at: http://www.iea.org/textbase/nppdf/free/2010/key_stats_2010.pdf. Accessed Sept 5 2010.

Judt, T. (2010). *Ill Fares the Land*. New York: Penguin.

Kelly, A. and D. Schindler, P. Hodson, J. Short, R. Radmonovitch, C. Nielsen (2010) Oil sands development contributes elements toxic at low concentrations to the Athabasca River and its tributaries. *Proceedings of the National Academy of Sciences. 107* (37), 16178–16183. http://www.pnas.org/content/107/37/16178.short. Accessed January 13 2011.

Levant, E. (2010) Ethical Oil: The case for Canada's Oil Sands. Toronto: McClelland and Stewart.

Meyer, D.S. & Staggenborg, S. (1996). "Movements, Countermovements, and the Structure of Political Opportunity." *American Journal of Sociology* 101(6):1628–1660.

Patels, W. (2010). Let's boycott all Greenies (and California). Global Post, July 15. Available at: http://www.globalpost.com/webblog/canada/lets-boycott-all-greenies-and-california. Accessed Dec 20 2010.

Pembina Institute. (2010). http://www.pembina.org/ Accessed 27 December 2010.

Perlin, J. (1989). *A Forest Journey: The Role of Wood in the Development of Civilization*. Cambridge: Harvard University.

Shove, E. (1997). Revealing the Invisible: sociology, energy, and the environment. In Redclift, M., & Woodgate, G. (Eds.). *The International Handbook of Environmental Sociology*, 261–273. Northhampton, Mass.: Edward Elgar.

Shove, E. & Warde, A. (1998). Inconspicuous Consumption: The Sociology of Consumption and the Environment. Department of Sociology, Lancaster University: http://www.lancs.ac.uk/fss/sociology/papers/shove-warde-inconspicuous-consumption.pdf.

Shove, E. & Walker, G. (2010). Governing transitions in the sustainability of everyday life. *Research Policy 39*, 471–476.

Sweeny, A. (2010). *Black Bonanza: Canada's Oil Sands and the Race to Secure North America's Energy Future*. Mississauga: Wiley.

Urry, J. (2010). Consuming the planet to excess. *Theory, Culture, Society, 27* (2–3), 191–212.

Zalik, A. (2010). Oil 'futures': Shell's Scenarios and the social constitution of the global oil market. *Geoforum, 41*, 553–564

Appendix

Still Photography Galleries of the Tar Sands Accessible Online

Louis Helbig's Aerial Photography images of the Athabasca tar sands: http://www.beautifuldestruction.ca/ and his stock image sheets of the region at http://www.beautifuldestruction.ca/stockcontactshee.html

Jiri Rezac Photographer: http://archive.jirirezac.com/

National Geographic Oil Sands Images (Photo's by Peter Essick): http://ngm.nationalgeographic.com/2009/03/canadian-oil-sands/essick-photography

Photographer Peter Essick's complete set of oilsands images (and pithy captions) at Aurora Photos: http://www.auroraphotos.com/SwishSearch?Keywords=essick+mcmurray&submit=Go!

Don Van Hout Athabasca River Expedition Connecting the Drops: http://www.connectingthedrops.ca/photos

Glenbow Museum Photograph Collection, Calgary, AB: http://ww2.glenbow.org/search/archivesPhotosSearch.aspx

Provincial Archives of Alberta Multimedia: https://hermis.alberta.ca/paa/Default.aspx?CollectionID=20

Oilsands Truth – Images by Project: http://oilsandstruth.org/index.php?q=tar-sands-photo-albums-project

Syncrude Canada Image Library: http://www.syncrude.ca/users/folder.asp?FolderID=5703

Petropolis Greenpeace Film: http://www.petropolis-film.com/

Greenpeace Canada Tar Sands Campaign Images: http://www.greenpeace.org/canada/en/campaigns/tarsands/Photos/

D.J. Davidson and M. Gismondi, *Challenging Legitimacy at the Precipice of Energy Calamity*, DOI 10.1007/978-1-4614-0287-9, © Springer Science+Business Media, LLC 2011

Getty Images (search for tars sands Fort McMurray):
http://www.gettyimages.ca/

Magnum images – advanced search by photographer

- See Jonas Bendiksen – Fort McMurray 2007 Album:
  http://www.magnumphotos.com/C.aspx?VP3=ViewBox&ALID=2K7O3RHKT
  7A2&IT=ThumbImage01_VForm&CT=Album
- Alex Webb – Fort McMurray Oilsands Album:
  http://www.magnumphotos.com/C.aspx?VP3=ViewBox&ALID=2K7O3R1VT
  X7L&IT=ThumbImage01_VForm&CT=Album

The Nature of Things – David Suzuki. Tipping Point:
Alberta's Oil Sands – January 27, 2011
http://www.cbc.ca/video/#/Shows/The_Nature_of_Things/1242300217/
ID=1769597772

Index